Scientific Workshop on Nuclear Fission and the Emission of Prompt Neutrons and Gamma Rays (Theory-5)

EPJ Web of Conferences Volume 256 (2021)

Barga, Italy
24-26 September 2019

Editors:

Nicolae Carjan
Stephan Oberstedt

ISBN: 978-1-7138-4179-1

Printed from e-media with permission by:

Curran Associates, Inc.
57 Morehouse Lane
Red Hook, NY 12571

Some format issues inherent in the e-media version may also appear in this print version.

For additional information, please contact EDP Sciences – Web of Conferences
at the address below.

EDP Sciences – Web of Conferences
17, Avenue du Hoggar
Parc d'Activité de Courtabœuf
BP 112
F-91944 Les Ulis Cedex A
France

Phone: +33 (0) 1 69 18 75 75
Fax: +33 (0) 1 69 28 84 91

contact-edps@webofconferences.org

Additional copies of this publication are available from:

Curran Associates, Inc.
57 Morehouse Lane
Red Hook, NY 12571 USA
Phone: 845-758-0400
Fax: 845-758-2633
Email: curran@proceedings.com
Web: www.proceedings.com

TABLE OF CONTENTS

edp sciences

EPJ Web of Conferences

Statement of Peer review

In submitting conference proceedings to EPJ Web of Conferences, I certify to the Publisher that I adhere to the **Policy on Publishing Integrity** of the journal in order to safeguard good scientific practice in publishing.

1. All articles have been subjected to peer review administered by the proceedings editors.
2. Reviews have been conducted by expert referees, who have been requested to provide unbiased and constructive comments aimed, whenever possible, at improving the work.
3. Proceedings editors have taken all reasonable steps to ensure the quality of the materials they publish and their decision to accept or reject a paper for publication has been based only on the merits of the work and the relevance to the journal.

Title, date and place of the conference

THEORY-5 Scientific Workshop on Nuclear Fission Dynamics and the Emission of Prompt Neutrons and Gamma Rays
Castelvecchio Pascoli, Barga (IT), September 24-26, 2019

Proceedings editor(s): Nicolae Carjan, Stephan Oberstedt

Date and editor's signature

Foreword

On September 24–26, 2019 a Scientific Workshop on Nuclear Fission Dynamics and the Emission of Prompt Neutrons and Gamma Rays (Theory-5) was held in Castelvecchio Pascoli, Barga, Italy. The workshop was organized by Unit G.2 Standards for Nuclear Safety, Security and Safeguards of the Joint Research Center (JRC), Geel, Belgium and the National Institute of Physics and Nuclear Engineering "Horia Hulubei", Bucharest, Romania. It was the fifth workshop of a series initiated in 2010 (Sinaia, Romania) and continued in 2012 (Biarritz, France), 2014 (Opatija, Croatia) and 2017 (Varna, Bulgaria).

The organizers were assisted by a Programme Advisory Committee consisting of Roberto Capote: IAEA, Austria, Diane Dore: CEA Saclay, France, Jovana Nikolov: University of Novi Sad, Serbia, Andreas Oberstedt: ELI-NP, Romania, Karl-Heinz Schmidt: GEF, Erzhausen, Germany, and Patrick Talou: LANL, USA, Anabella Tudora: Bucharest University, Romania.

The purpose of the workshop was to provide a comprehensive overview of the dynamical aspects of nuclear fission and their consequences on fission fragment properties. The contributions have highlighted recent theoretical advancements and precise measurements performed at state-of-the-art facilities around the world. The success of the workshop was due to the 33 carefully prepared contributions and to the lively discussions they stimulated.

We would like to dedicate this volume to the memory of Dr. Mihai Mirea who passed away on August 27, 2020. He was an extremely valuable participant at all five Theory-X workshops. We will certainly miss him.

A very special 'thank you' goes to the perfect local organizational support of Mira Van de Lucht.

Stephan Oberstedt and Nicolae Carjan

October 2021

Isomer yields in nuclear fission

A. Al-Adili[1,*], *Z.* Gao[1], *M.* Lantz[1], *A.* Solders[1], *M.* Österlund[1], and *S.* Pomp[1]

[1]Department of Physics and Astronomy, Uppsala University, Uppsala, Sweden

Abstract. The generation of angular momentum in the fission process is still an open question. To shed light on this topic, we started a series of measurements at the IGISOL-JYFLTRAP facility in Finland. High-precision measurements of isomeric yield ratios (IYR) are performed with a Penning trap, partly with the aim to extract average root-mean-square (rms) quantities of fragment spin distributions. The newly installed Phase-Imaging Ion-Cyclotron Resonance (PI-ICR) technique allows the separation of masses down to tens of keV, which is sufficient to disentangle many isomers. In this paper, we first summarize the previous measurements on the neutron and proton-induced fission of uranium and thorium, e.g. the odd cadmium and indium isotopes ($119 \leq A \leq 127$). The measurements revealed systematic trends as function of mass number, which stimulated further exploration. A recent measurement was performed at IGISIOL and several new IYR data will soon be published, for the first time. Secondly, we employ the TALYS nuclear-reaction code to model one of the newly measured isomer yields. Detailed GEF and TALYS calculations are discussed for the fragment angular momentum distribution in ^{134}I.

1 Introduction

While isomeric fission yields are an important part of a full fission-process description, the data libraries are incomplete and lack important systematic investigations of the isomer production cross section. Often isomeric ratio data are originating from γ-ray measurements, and there are wide discrepancies in the reported yields, most likely due to lack of nuclear structure data. Novel methods that reduce the dependence on level structures are important to improve the isomeric yield data situation. One example is the direct ion counting method used at IGISOL in Jyväskylä, Finland [1, 2]. High precision mass measurements are performed using a Penning trap, which allows the mass separation of a meta-stable state from the ground state down to a mass difference of 50 keV [3].

The fission fragments (FF) typically carry large angular momenta, J. The generation of J still comprises challenges for contemporary fission modeling [4]. Many questions are unanswered, e.g. the conflicting experimental results concerning a possible saw-tooth trend in the average $J(A)$ [5]. Another important question is how angular momentum is shared among individual and collective degrees of freedom and how that eventually depends on the compound nucleus and excitation energy.

With the aid of nuclear de-excitation codes, one can model the angular momentum population of the primary fission fragments [6]. Albeit strongly model-dependent, this can shed

*e-mail: ali.al-adili@physics.uu.se

some light on the systematics of angular momentum generation and the dependency on nuclear shell structure. The initial angular momentum possessed in the fragments will influence the number of emitted particles and their energy spectra. One such method was developed utilising the TALYS 1.9 reaction code [7]. Detailed description and benchmarking of this de-excitation model was published elsewhere [8].

2 Summary of IYR experiments at IGISOL

Fission fragments are produced by light charged particles accelerated up to tens of MeV, that impinge on a fissionable target material [9]. Fission products stop in a helium-filled gas cell and are extracted for mass identification. Detailed simulations of the gas cell stopping efficiency and the associated systematic ion counting uncertainties can be found elsewhere [10]. A first isobaric selection (A/q) is performed in a dipole magnet and a subsequent Paul trap cools and bunches the ions. The ion bunches are then injected into the Penning trap which consists of a purification trap and a precision trap [9].

Five IYR experiment campaigns have been performed at IGISOL within this collaboration, one on ^{nat}U(n,f), three on ^{nat}U(p,f) and one on ^{232}Th(p,f). Three different methods have been utilized for particle identification and counting: γ-ray spectroscopy, side-band cooling technique [9] and Phase-Imaging Ion-Cyclotron Resonance (PI-ICR) technique [11]. A summary of the measurements and the associated analysis methods is extensively covered elsewhere [12, 13]. Here we review the most important findings from each measurement campaign.

2.1 IYRs from ^{nat}U(n,f)

The first isomeric yield ratios measured within the Uppsala-Jyväskylä collaboration was based on γ-ray spectroscopy using High-Purity Germanium detectors. Yield ratios of three tin isotopes (cumulative) and two antimony isotopes (independent) were measured in the ^{nat}U(n,f) reaction with neutron energies of $E_n = 12.4 \pm 8.8$ MeV [12, 14]. The results were suffering from rather large uncertainties due to low neutron production, but still revealed some interesting differences to evaluated ENDF B-VII.1 files.

2.2 IYRs from ^{nat}U(p,f) and ^{232}Th(p,f) with Side-band cooling

In the second campaign the focus was put on using the Penning trap and to measure several new isomeric yields using the so-called side band cooling technique [1]. The studied reactions were ^{nat}U(p,f) and ^{232}Th(p,f) with $E_p = 25.0$ MeV. Results for six fission products (three from the light fragment peak and three from the heavy peak) were reported [1]. An example is shown in Fig. 1 for ^{81}Ge. The plot shows a frequency scan in the Penning trap, which upon matching a characterising resonance frequency, filters the ions (based on mass). Differences between the two fissioning systems were observed, hinting to differences in populated angular momenta. An attempt to extract angular momentum root-mean-square (rms) values J_{rms} was done. Most deduced values turned out to be much larger than what is observed in thermal reactions, hinting to a larger angular momentum population in the compound nucleus at higher excitation energies [15].

2.3 IYRs from ^{nat}U(p,f) with PI-ICR

The side-band cooling technique has a mass resolution limited to about half an MeV, which is illustrated in the example of ^{81}Ge (Fig. 1). The mass difference between ground- and excited

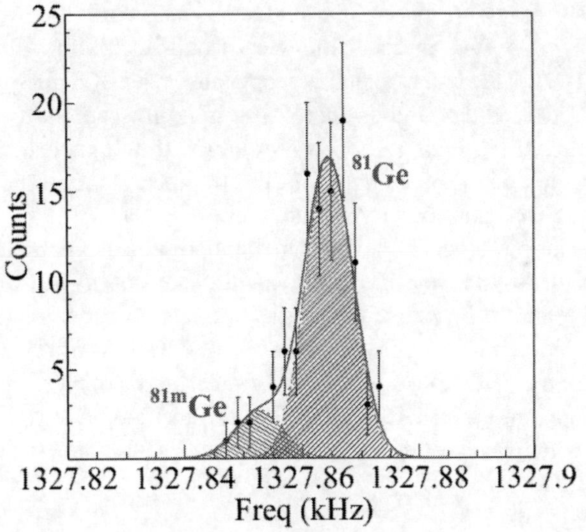

Figure 1. The IYR of ^{81}Ge extracted through Gaussian fits to the isomer (E = 679 keV) and the ground state [1]. The data were obtained using the side-band cooling technique, which has a mass resolution limit of about 500 keV for isomer separation.

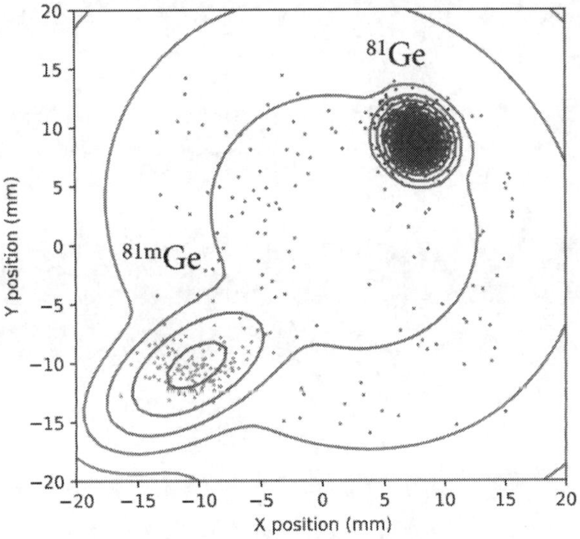

Figure 2. An example from ^{81}Ge, where the Phase-Imaging Ion-Cyclotron Resonance (PI-ICR) technique was employed. The plot illustrates the X,Y coordinates of a position-sensitive Micro-Channel Plate. The ground state is well separated from the meta-stable state.

states in ^{81}Ge is 679 keV, which causes the peaks to overlap. This leads to growing uncertainties in the deduced IYR. With the birth of the novel PI-ICR technique [3], the achieved mass resolution is about 10 times better, proving the ^{81}Ge isomers easy to disentangle as seen in Fig. 2 [13]. A two-dimensional fit is performed using Bayesian mixing which yields the weight of each isotope by calculating the probability of each data point belonging to any of the peaks.

The new PI-ICR technique has been used in two measurement campaigns on $^{nat}U(p,f)$ and with 25 MeV protons. A systematic study was made on 10 new isomeric yields, five odd-A cadmium isotopes and an equal number of odd-A indium isotopes, all produced in the fission of $^{nat}U(p,f)$. The data revealed a decreasing trend of high-spin state population as a function of mass for indium, and an increasing trend for cadmium [2]. After calculating the root-mean-square of J, a correlation was found with the nuclear quadrupole moments. A possible explanation could be a change in the fragment shape which causes the angular momentum to change, because of varying moment of inertia.

A second campaign was recently completed and 18 new isomeric ratios were measured, close to shell closure in the Sn-132 region. A total of about 30 isomers have now been measured, with the aim to investigate how shell effects influence the IYR and the angular momentum population. Analysis of the most recent data is still ongoing. Below we show detailed de-excitation calculations for one of the measured isotopes, ^{134}I, to demonstrate the most recent development in our GEF+TALYS-based method. A simplified decay scheme is depicted in Fig. 3 which involves the lower discrete levels including the meta-stable state (8^-) and the ground state (4^+).

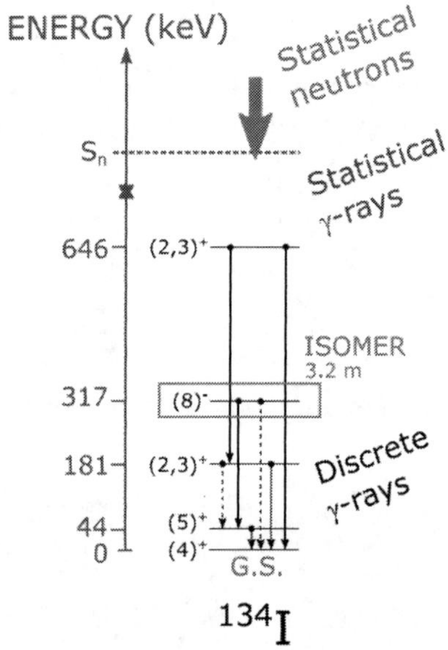

Figure 3. A simplified decay scheme of ^{134}I. The isomeric state has an energy of 317 keV and decays by internal transition.

3 Model calculations for ^{134}I

TALYS uses the statistical Hauser-Feshbach evaporation model which takes into account the competition between neutrons and γ-rays, in addition to the RIPL-3 data-base for discrete levels. We have used the Microscopic-Goriely level density model and up to 10 discrete levels for each nucleus. In order to perform the calculations, one needs the initial parameterisation of excitation energy and angular momentum. The spin distribution is assumed to follow the commonly used level density function, based on a Rayleigh probability distribution:

$$P(J) \propto (2J+1)\exp\left(-\frac{(J+0.5)^2}{B^2}\right). \tag{1}$$

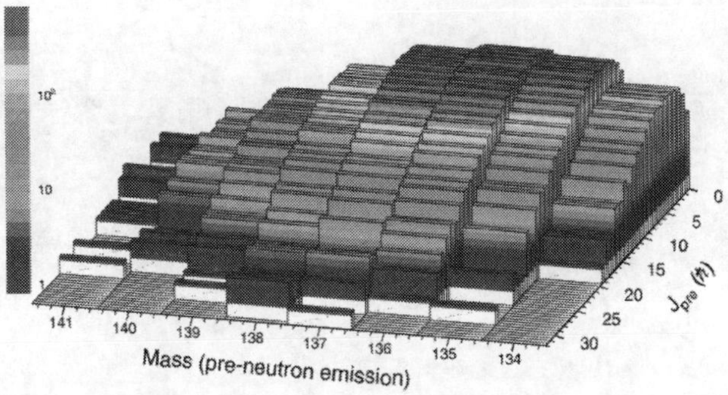

Figure 4. The angular momentum distributions obtained with GEF, for each fission fragment leading to ^{134}I as fission product after neutron emission. Each fission fragment case was modelled in TALYS for different values of B (see Eq. (1)).

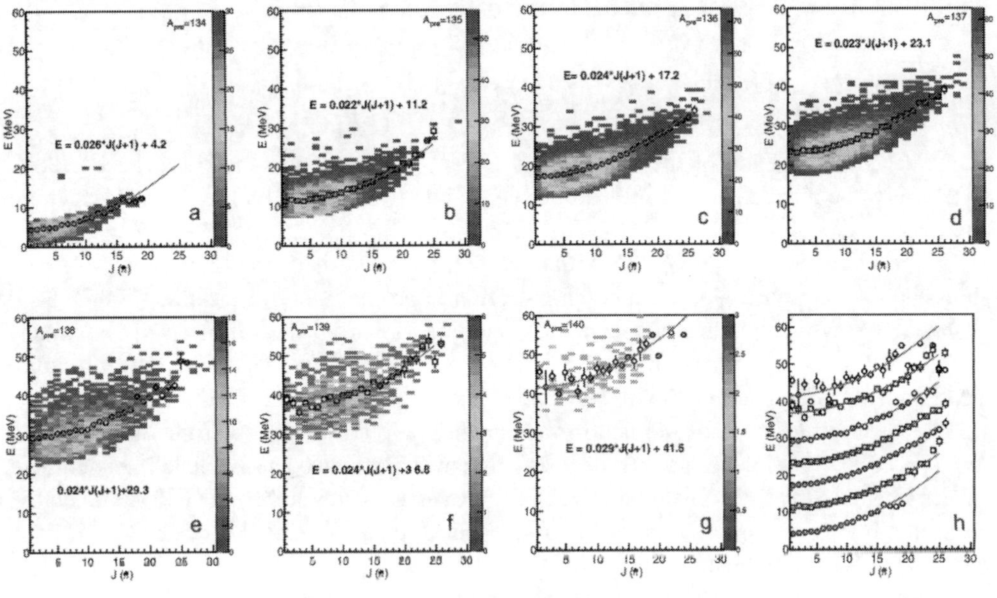

Figure 5. The excitation energy, E, versus spin, J, for all fission fragments leading to ^{134}I, as simulated with GEF. The average energies are parameterised by Eq. (2) with C_1 and C_2 acting as free fit parameters. Plot h shows all the fits from a–g.

The "spin cut-off" parameter B, defines the mean and width of the distribution and is normally approximated with the root-mean-square, J_{rms}, of $P(J)$. By varying B, it is possible to obtain the value that reproduces the experimental result [8]. In this work we use the prediction IYR value of GEF as the benchmark for these calculations.

The angular momentum distributions were simulated using the GEF code (V.2017-1.2) [16], for all fragments leading to ^{134}I after neutron emission (Fig. 4). Up to 7 neutrons may be emitted, all with a unique $P(J)$ distribution. Figure 5 illustrates the excitation energy versus spin for each fission fragment leading to ^{134}I. The average excitation energy, $\bar{E}$, increases as

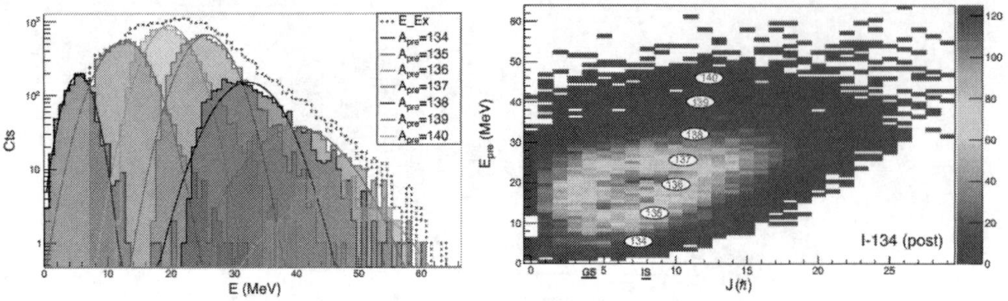

Figure 6. *Left*: the excitation energy distribution for each fission fragment leading to ^{134}I after neutron emission. Normal distributions were fitted to each distribution to retrieve average energies and spreads $\sigma_{\rm E}$. *Right*: the spin versus E^* distribution for all distributions in Fig. 5, superimposed. The spin values of ground state and meta-stable state are indicated on the spin axis. The average excitation energies (left plot of this figure) and $J_{\rm rms}$ values are indicated for each individual fragment by numbered circles. The total GEF $P(J)$ distribution has a $J_{\rm rms}$ value of 9.5 $\hbar$.

function of fragment spin J due to the increase in rotational energy:

$$\bar{E}(J) = C_1\left(J^2 + J\right) + C_2 \propto \frac{\hbar^2}{2\mathcal{J}} J\,(J+1)\,. \tag{2}$$

where $\mathcal{J}$ is the moment of inertia. C_1 and C_2 act as free fit parameters. The coefficient C_1 seems to be independent of the parent mass while C_2 depends on the neutron separation energy.

In this work, we provide TALYS with input files parameterising the excitation energy with Gaussian distributions, which is a good first order approximation for each neutron emission channel. We deduce the average energy and spread of energy ($\sigma_{\rm E}$) from each Gaussian fit as seen in Fig. 6 (left). More refined excitation energy parameterisation is currently under development, where the functional dependencies of Eq. (2) and Fig. 5 will be implemented.

The superposition of all the neutron channels is plotted in Fig. 6 (right) together with the respective average energies (from the Gaussian fits) and $J_{\rm rms}$ as indicated by numbered circles (mass number, A). The total angular momentum distribution from GEF, summed over all fission fragments, has a $J_{\rm rms}$ of 9.5 $\hbar$. As expected, there is a positive correlation between the average energy and the average spin [13].

All in all, 56 TALYS IYR calculations were performed (eight spin distributions for each of the seven pre-cursors), all of which are plotted in Fig. 7. The IYR is defined as the ratio of high-spin cross section to the total production cross section. As the B parameter increases, higher spin values are populated by the de-exciting nucleus. Therefore, the population of the high-spin state will increase with B. The vertical lines correspond to $J_{\rm rms}$ as extracted from the GEF distributions plotted in Fig. 4. GEF predicts an IYR of about 0.55. Based on this value the average $J_{\rm rms}$ from TALYS is between 9.5 and 10.5 $\hbar$, for the case of ^{134}I.

4 Conclusions and outlook

With the different techniques employed, we have thus far measured IYR for 34 different products for three different reactions at the IGISOL facility in Finland [1, 2, 14]. The bulk of the data, 30 IYR, are in the ^{132}Sn region and for ^{238}U(p,f) at 25 MeV. The analysis carried out on the data this far have already revealed interesting trends. It remains to be seen what can be learned from the most recent data-set, offering to study the behaviour across the $N = 82$ line

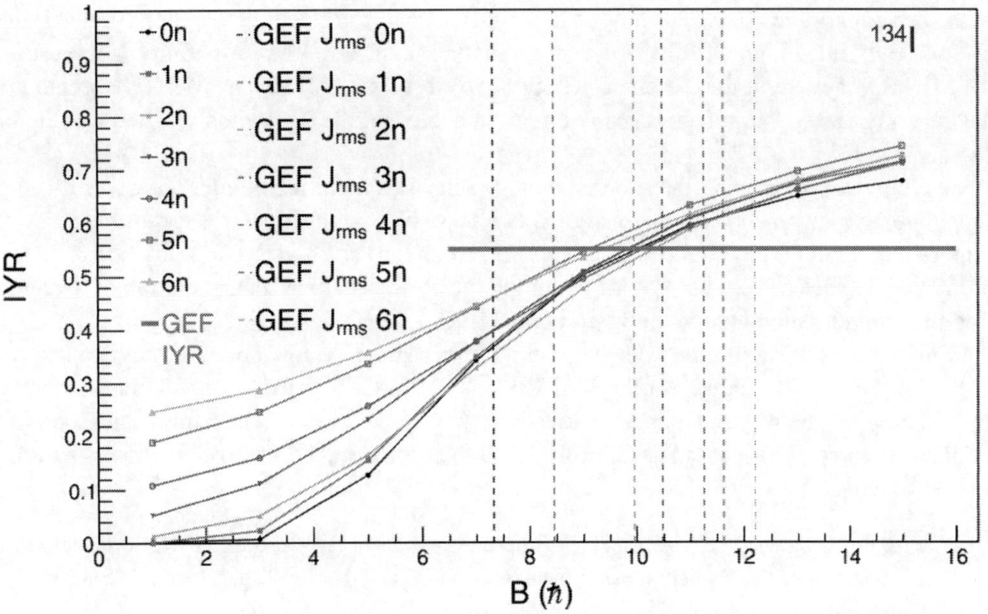

Figure 7. TALYS and GEF calculations for the isomeric yield ratio of ^{134}I, for all fragments ($134 \leq A \leq 140$u) giving rise to ^{134}I, after prompt neutron emission. The deduced GEF J_{rms} values are indicated by the vertical lines for each neutron channel. GEF gives a total of 9.5 $\hbar$ as an average of all channels and the TALYS calculations point to a J_{rms} value between 9.5 and 10 $\hbar$.

for three different elements, as well as trends along $N = 81$. There are several different options that seem worth exploring and that will strengthen the experimental database to improve our understanding of the fission process and to develop and constrain fission model codes.

One option is to study the obtained cadmium and indium behavior of the IYR in the ^{238}U(n,f) case at lower excitation energies to allow unfolding the contributions from different fission systems by reducing the contributions from higher fission chances. Measuring at, say, 12 MeV, also offers the chance to see possible effects from the angular momentum of the compound system.

Another option is to expand the studies to other regions of the nuclear chart. Especially interesting would be to look at the ^{78}Ni region and to explore complementary fission fragments. While the technique does not allow for coincidence measurements, it might be fruitful to explore trends along complementary nuclear charges, e.g., compare the trends seen for In nuclei ($Z = 49$), with trends at $Z = 44$ (Tc) in the U(p,f) case, or $Z = 42$ (Mo) for proton-induced-fission of thorium.

Thirdly, it is of course desirable to expand the IYR database for the two systems were we thus far only scratched the surface, as well as studying other systems. Particularly interesting would be the case of spontaneous fission, e.g., of ^{248}Cm or ^{252}Cf, without the need for an accelerator.

To achieve these ends, there are some possible developments at the IGISOL facility. A development of new ion guides are important, where we try to increase stopping efficiencies and decrease extraction times. This could facilitate more measurements of IYR in neutron-induced fission.

Using other measurement techniques to measure IYR that are not measurable at JYFLTRAP, should also be looked at. While the PI-ICR technique offers excellent mass resolving power (MRP), MR-TOF systems allow for faster measuring times. Thus, isotopes with shorter half-lives as well as lower yield regions can be targeted. One option for this is the use of newly implemented MR-TOF system in Jyväskylä. Another possibility is the use of the MR-TOF system at the Cryogenic Stopping Cell (CSC) at GSI. The MR-TOF technique has recently showed significant improvement in achievable MRP, which reached a value of 10^6 at GSI [17].

We also consider using the more standard γ-measurement techniques for cases of very short half-lives or low excitation energies, not accessible with the above techniques, but of importance for our study. Following along this research track should result in a rich database of IYR for various cases, and provide insight in the physics behind the generation of the large fragment angular momenta in the fission process.

Finally, we envisage further development of the TALYS-driven model through collaboration with the code developers at the IAEA. Plans for further verification and benchmarking are ongoing, especially to understand how other fission reaction mechanisms (e.g. Coulomb excitation, abrasion fission and fusion-fission) affect the de-excitation of fragments and the isomer population.

Funding for this work was received from the European Commission within the CHANDA project (contract 605203), the Swedish Radiation Safety Authority (SSM) and the Swedish Research Council (VR).

References

[1] V. Rakopoulos, M. Lantz, A. Solders, A. Al-Adili, A. Mattera, L. Canete, T. Eronen, D. Gorelov, A. Jokinen, A. Kankainen et al., Phys. Rev. C **98**, 024612 (2018)
[2] V. Rakopoulos, M. Lantz, S. Pomp, A. Solders, A. Al-Adili, L. Canete, T. Eronen, A. Jokinen, A. Kankainen, A. Mattera et al., Phys. Rev. C **99**, 014617 (2019)
[3] S. Eliseev, K. Blaum, M. Block, C. Droese, M. Goncharov, E. Minaya Ramirez, D.A. Nesterenko, Y.N. Novikov, L. Schweikhard, Phys. Rev. Lett. **110**, 082501 (2013)
[4] A.N. Andreyev, K. Nishio, K.H. Schmidt, Rep. Prog. Phys. **81**, 016301 (2018)
[5] C. Wagemans, *The nuclear fission process* (CRC Press, 1991), ISBN 978-0849354342
[6] S. Okumura, T. Kawano, P. Jaffke, P. Talou, S. Chiba, J. Nucl. Sci. Technol. **55**, 1009 (2018)
[7] A. Koning, D. Rochman, S. van der Marck, Nucl. Data Sheets **118**, 187 (2014)
[8] Al-Adili, A., Rakopoulos, V., Solders, A., Eur. Phys. J. A **55**, 61 (2019)
[9] H. Penttilä, P. Karvonen, T. Eronen, V.V. Elomaa, U. Hager, J. Hakala, A. Jokinen, A. Kankainen, I.D. Moore, K. Peräjärvi et al., Eur. Phys. J. A **44**, 147 (2010)
[10] A. Al-Adili, K. Jansson, M. Lantz, A. Solders, D. Gorelov, C. Gustavsson, A. Mattera, I. Moore, A.V. Prokofiev, V. Rakopoulos et al., Eur. Phys. J. A **51**, 59 (2015)
[11] D.A. Nesterenko, T. Eronen, A. Kankainen, L. Canete, A. Jokinen, I.D. Moore, H. Penttilä, S. Rinta-Antila, A. de Roubin, M. Vilen, Eur. Phys. J. A **54**, 154 (2018)
[12] A. Mattera, Ph.D. thesis, Uppsala University, Applied Nuclear Physics (2017)
[13] V. Rakopoulos, Ph.D. thesis, Uppsala University, Applied Nuclear Physics (2018)
[14] A. Mattera, S. Pomp, M. Lantz, V. Rakopoulos, A. Solders, A. Al-Adili, H. Penttilä, I. D. Moore, S. Rinta-Antila, T. Eronen et al., Eur. Phys. J. A **54**, 33 (2018)
[15] D.C. Aumann, W. Gückel, E. Nirschl, H. Zeising, Phys. Rev. C **16**, 254 (1977)
[16] K.H. Schmidt, B. Jurado, C. Amouroux, C. Schmitt, Nucl. Data Sheets **131**, 107 (2016), special Issue on Nuclear Reaction Data
[17] T. Dickel, Priv. Comm. (2019)

Angular distribution of fragments in neutron-induced nuclear fission at energies 1–200 MeV: Data, theoretical models and relevant problems

Alexey Barabanov[1,*], *Alexander* Vorobyev[2], *Alexei* Gagarski[2], and *Oleg* Shcherbakov[2], and *Larisa* Vaishnene[2]

[1]NRC Kurchatov Institute, 123182 Moscow, Russia
[2]NRC Kurchatov Institute, B.P. Konstantinov Petersburg Nuclear Physics Institute, 188300 Gatchina, Russia

Abstract. In recent years, investigations of angular distributions of fragments in neutron-induced nuclear fission have been extended to intermediate energies, up to 200 MeV, as well as to a wide range of target isotopes. Using as an example the latest data obtained by our group for the reaction ^{237}Np(n, f), we discuss the specific features of fission fragment angular distribution and present a method for their simulation based on the code TALYS. It is shown that a simplified model reasonably describes energy dependence of the angular distribution in the whole range 1–200 MeV. The ways to improve the model are discussed along with the possibilities to use it for obtaining new information on fission and pre-equilibrium processes in neutron-nucleus interaction. We consider also the relevant problems of describing fission fragment angular distributions.

1 Introduction

Within the last five years we have studied angular distributions of fragments in neutron induced fission at energies 1–200 MeV for the following isotopes: ^{232}Th, ^{235}U, ^{238}U [1], ^{209}Bi, ^{233}U [2], ^{nat}Pb, ^{239}Pu [3], and ^{237}Np [4] (see also [5, 6]). Measurements were made with the use of the pulsed spallation neutron source and neutron TOF spectrometer GNEIS based on the 1 GeV proton synchrocyclotron in the Petersburg Nuclear Physics Institute [7, 8]. Similar investigations were performed also by collaborations n_TOF and NEFFTE for target nuclei ^{232}Th [9], ^{235}U, ^{238}U [10] and ^{235}U [11], respectively. These studies are of significance because almost all earlier obtained data on fission fragment angular distributions refer to the energy region below 20 MeV; only in the work [12] the angular anisotropy of fragments in neutron induced fission of ^{232}Th and ^{238}U was measured up to the energy 100 MeV.

These studies using neutrons of intermediate energies of up to 200 MeV and higher are important for the development of new nuclear technologies, such as nuclear power based on ADS (Accelerator-Driven Systems), transmutation of nuclear waste, radiation testing of materials, nuclear medicine. Besides, the data on fission fragment angular distributions are of interest for nuclear science by stimulating the progress of theoretical models for neutron-nucleus collisions and accompanying processes. Fission is among the most noticeable ones

*e-mail: a_l_barabanov@mail.ru

if one uses heavy target nuclei. Moreover, the fission is of special interest due to the fact that a nucleus undergoes fission while being in the equilibrium excited state only, i.e. when the excitation enegy is distributed over all degrees of freedom, including collective ones. That is why at high collision energy, when pre-equilibrium processes play substantial role both in the reaction and subsequent nuclear decay, restrictions on their contributions may be obtained from the observation of fission.

Many other specific features of fission including fission fragment angular anisotropy are related to transition states on barriers first introduced by A. Bohr [13]. The transition state is characterized by a projection K of the nuclear spin J on the deformation axis which transforms into the direction of fission axis. Thus the quantum number K along with a spin projection M on the z axis of laboratory reference system determines an angular distribution of fission fragments $dw^J_{MK}(\omega)/d\Omega$. Therefore the angular distribution of fragments is among observables, which are the most sensitive to the transition states characteristics, in particular, to the K value. This circumstance resulted in intensive experimental studies of fission fragment angular anisotropy during 2–3 decades after discovery of this phenomenon in the energy region below 20 MeV available at that time (see, e.g., [14]).

In this work we discuss both the total fission cross section σ_f and the fission fragment angular distribution,

$$W(\theta) \equiv \frac{dw(\theta)}{d\Omega} = \frac{1}{\sigma_f}\frac{d\sigma_f(\theta)}{d\Omega}, \tag{1}$$

which is a ratio of differential $d\sigma_f(\theta)/d\Omega$ to total σ_f fission cross sections, in the wide range 1–200 MeV of neutron energy. Here θ is the angle between the light (for definiteness) fragment momentum and the momentum of incident neutrons, taken for the axis z. The possibilities to obtain new information on fission and pre-equilibrium processes in neutron-nucleus interaction from the specific features of $W(\theta)$ are of main interest. We present a model describing energy dependence of the angular distribution. Throughout the work we use as an example the data for the reaction ^{237}Np(n, f). We consider also the relevant problems in this field and perspectives for the development.

2 Total fission cross section

Currently, there are a number of computer codes focused on modeling nuclear reactions in a wide energy range, covering the interval of 1–200 MeV; among them the nuclear reaction program TALYS [15]. However, TALYS as well as the other known similar programs has a limited functional for describing differential cross sections, in particular, it does not compute differential fission cross section. Let us consider shortly the main features of the TALYS program. It deals only with binary channels, like many other similar codes. Thus, at the first stage of reaction, just after neutron-nucleus collision, only two possibilities are taken into account: either a compound nucleus is formed (let it consists of Z_0 protons and N_0 neutrons) and then decays into one of the binary channels (fission is among them), or a particle (γ, n, p, d, t, h, α) and corresponding residual nucleus (which is generally excited) are produced in a direct or pre-equilibrium process. Thus, the cross section of fission at the first stage of the reaction (binary fission in terms of the TALYS Manual) is given by

$$\sigma_f^{(1)} = \sum_{J\pi} \sigma_{Z_0N_0}(J\pi)\, P_f^{Z_0N_0}(J\pi), \tag{2}$$

where $\sigma_{Z_0N_0}(J\pi)$ and $P_f^{Z_0N_0}(J\pi)$ are the population cross section and the fission probability (fissibility) of the compound state with spin J and parity π. However, this is only a part of

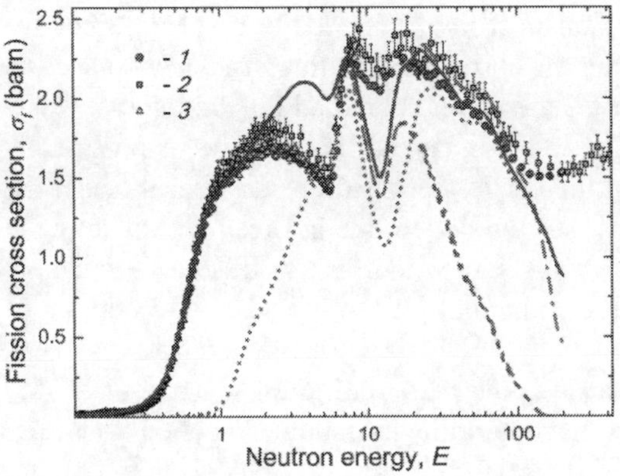

Figure 1. Fission cross section for ^{237}Np versus neutron energy E. Experimental data: 1 – [17], 2 – [18], 3 – [19]. Dotted line corresponds to calculation with the default TALYS parameters. Other lines are obtained with the corrected parameters (see text): solid and dot-dashed lines correspond to σ_f with and without account for multiple pre-equilibrium emission, dashed and dot-dot-dashed lines correspond to σ_f^C with and without account for multiple pre-equilibrium emission.

the total fission cross section because any excited residual nucleus formed at any stage of the reaction may undergo fission.

Thus, at the first reaction stage residual nuclei arise either in the primary compound nucleus decay or as the result of a direct or pre-equilibrium process. In the first case, the residual nucleus is assumed to be in an equilibrium excited state (in other words, to be a secondary compound nucleus); it decays at the second reaction stage into a particle and a new residual nucleus or divides into two fragments. In the second case, the residual nucleus is, generally, in a non-equilibrium excited state, therefore, it decays into a particle and a residual nucleus either due to a pre-equilibrium process (this is a multiple pre-equilibrium emission), or after transition to the compound state with an additional possibility to undergo fission. The same is true for third and subsequent reaction stages. Lets i numerates the levels of the nucleus (Z, N) with the same spin J and parity π (in the nucleus (Z_0, N_0) the number i_0 corresponds to the primary compound state). Then the population cross section $\sigma_{ZN}(J\pi i)$ of a level is a sum of two contributions,

$$\sigma_{ZN}(J\pi i) = \sigma_{ZN}^{DPE}(J\pi i) + \sigma_{ZN}^{C}(J\pi i), \tag{3}$$

where the first one corresponds to the case, when a direct (D) or pre-equilibrium (PE) process preceeds the level population, while the second one – to the case, when the level is populated as a result of statistical decay of the primary compound nucleus. The total fission cross section is given by

$$\sigma_f = \sum_{ZNJ\pi i} \sigma_{ZN}(J\pi i)\, P_f^{ZN}(J\pi i), \tag{4}$$

where $P_f^{ZN}(J\pi i)$ is the level fissibility, and summation over i be the integration when the levels belong a continuum (the cross Sect. 2 can be extracted from (4) keeping only the terms corresponding to $Z = Z_0$, $N = N_0$, $i = i_0$ with $\sigma_{Z_0N_0}(J\pi) = \sigma_{Z_0N_0}^{C}(J\pi i_0)$). We see that according (3) and (4) the total fission cross section divides into two parts, σ_f^{DPE} and

$$\sigma_f^C = \sum_{ZNJ\pi i} \sigma_{ZN}^{C}(J\pi i)\, P_f^{ZN}(J\pi i), \tag{5}$$

depending on either we deal with a residual nucleus fission preceeded by a direct or pre-equilibrium process, or with the primary compound nucleus fission or fission of a residual nucleus formed as a result of its statistical decay.

In the TALYS program, the fission cross Sect. 4 is calculated using a number of models, in particular, the optical one, as well as for direct and pre-equilibrium processes, level density of excited nuclei, probabilities of radiation transitions, nuclear fission. Each model implies many parameters. One can take the values that are considered as optimal and set by default, or change them. Many default parameters are taken from the RIPL library [16]. Since we deal with the reaction ^{237}Np(n, f), let us consider the corresponding total fission cross section measured in [17–19]. In the last of these works, the results obtained for the interval 0.1–9 MeV were compared with the calculated values. It turned out that the use of parameters from the RIPL library – heights and widths of the barriers, transition states, level density at the barriers – leads to significant discrepancies: see the dotted line in Fig. 1 obtained by the TALYS program with the default parameters. However, the modification of some parameters for the isotopes ^{238}Np, ^{237}Np, ^{236}Np allowed the authors of [19] to get a good description of the observed fission cross section.

One can expect that the angular distribution (1) as the ratio of differential and total fission cross sections is not sensitive to all parameters that determine fission cross section (these expectations are shown below to be justified, at least partially). But we should take into account the decisive role of transition states in forming angular distribution of fission fragments. That is why we restricted ourselves to correcting the heights and widths of the barriers, as well as the transition states for the isotopes ^{238}Np, ^{237}Np, ^{236}Np. Choosing the default option of explicit account of collective enhancement of the level density for nuclei at barriers, we obtained using the TALYS code the total fission cross section as function of the incident neutron energy, which reasonably agrees with the results of measurements in the range of 0.1–100 MeV. Noticeable discrepancies, not exceeding 25–30%, remain only in the range of 2–16 MeV and above 100 MeV (see Fig. 1 and comments in the next paragraph). To enhance the fission cross section near the thresholds of the 2nd and 3rd fission chances, the corrections we made are not enough; it is necessary to increase additionally the number of transition states (the level densities on the barriers) for the ^{238}Np and ^{237}Np isotopes. However, this is the factor which can be neglected according the above expectations. The region above 100 MeV is dicussed in Sect. 5.

The experimental data and the calculated curves are presented in Fig. 1. The calculations were performed with and without account for multiple pre-equilibrium emission (in ref. [4] we have considered only the second option). Both the total fission cross section (4) and its compound component σ_f^C (5) are shown in Fig. 1. For details about the used corrected parameters for barriers and transition states we refer to [4] (transition states correspond to variant 1 from this paper). The calculations of the fission fragment angular anisotropy for the reaction ^{237}Np(n, f) described below were performed with the same corrected parameters.

3 Fission fragment angular distribution

According to A. Bohr [13] the angular distribution of fragments from fission of transition state with spin J and its projections M and K to the z axis and deformation axis, respectively, is given by

$$\frac{dw^J_{MK}(\theta)}{d\Omega} = \frac{2J+1}{4\pi}\,|d^J_{MK}(\theta)|^2. \tag{6}$$

Indeed, the spatial part of the wave function of the axially symmetric nucleus at such transition state is proportional to the Wigner function $D^J_{MK}(\omega) = e^{iM\phi}\, d^J_{MK}(\theta)\, e^{iK\psi}$, where Euler angles $\omega = (\phi, \theta, \psi)$ determine the orientation of the reference frame, related to the nucleus, with respect to the laboratory system (x, y, z). Thus, the square of the modulus of the Wigner function determines the distribution in space of the direction of the deformation axis and, consequently, the direction of fission axis. Notice, that the spin of compound nucleus $\mathbf{J} = \mathbf{s}+\mathbf{I}+\mathbf{l}$ is combined of neutron spin $s = 1/2$, target-nucleus spin I, and neutron orbital momentum $\mathbf{l} = [\mathbf{r}\times\mathbf{k}]$, where $\mathbf{k}$ is the momentum of incident neutron. When s and I are not oriented, then the spin J is directed predominantly transversely to the z axis because $\mathbf{l} \perp \mathbf{k}$. In other words, the compound nucleus states with small projections M of spin J are predominantly populated, with states M and $-M$ are populated with equal probabilities (this type of spin orientation is referred as spin alignment). Spin orientation of fissioning compound state is a necessary condition for angular anisotropy of emitted fragments.

When the oriented compound nucleus decays into the particle and residual nucleus, then the spin orientation partially transfers to this nucleus. Thus, the fragments from its fission are emitted also anisotropically with respect to the z axis. The same is true for the subsequent stages of the primary compound nucleus decay. In a similar way, when a direct or pre-equilibrium process occurs at the first reaction stage, then, in principle, the formed residual nucleus also obtains some spin orientation. That is why the fission fragment angular anisotropy may arise when both this nucleus as well as the nuclei resulting from its decay undergo fission. Thus, the differential fission cross section (and the angular distribution of fission fragments) depends on the spin orientation of all states (with nonzero spins) of all nuclei forming in the reaction and capable for fission. This orientation, in turn, is determined by the population cross sections $\sigma_{ZN}(J\pi iM)$ for states with particular projections M of spin J on the z axis or, what is equivalent, by the diagonal elements of the spin density matrix

$$\eta_{ZN}^{J\pi i}(M) = \frac{\sigma_{ZN}(J\pi iM)}{\sigma_{ZN}(J\pi i)}, \qquad \sum_M \eta_{ZN}^{J\pi i}(M) = 1. \tag{7}$$

These quantities are not calculated in TALYS and other similar programs known to us, therefore, they do not have the option of computing the angular distribution of fragments (as well as the differential fision cross section).

In general, the angular distribution $dw_{ZN}^{J\pi i}(\theta)/d\Omega$ of fragments from fission of a nucleus (Z,N) at a level (J,π,i) is a superposition of the angular distributions (6) and has the form

$$\frac{dw_{ZN}^{J\pi i}(\theta)}{d\Omega} = \sum_M \eta_{ZN}^{J\pi i}(M) \sum_K \rho_{ZN}^{J\pi i}(K) \frac{dw_{MK}^{J}(\theta)}{d\Omega}, \tag{8}$$

where the probability distribution over the number K is introduced

$$\rho_{ZN}^{J\pi i}(K) = \frac{P_f^{ZN}(J\pi iK)}{P_f^{ZN}(J\pi i)}, \qquad \sum_K \rho_{ZN}^{J\pi i}(K) = 1. \tag{9}$$

Here, $P_f^{ZN}(J\pi iK)$ is the fission probability via the transitional states with the projection K of spin J on the deformation axis (note that variants of expression (8) are sometimes called as the Halpern–Strutinsky formula – see, e.g., [20]). The distribution (9) is determined by the fission mechanism (in particular, the nucleus may not have axial symmetry on the barrier; in this case, some mixing over K is known to occur). It is important that $\rho^{J\pi}(K) = \rho^{J\pi}(-K)$ in the absence of parity violation effects (they are very small and usually may be neglected). Thus, if the angular distributions (8) normalized to unity are known, then the differential fission cross section can be written as

$$\frac{d\sigma_f(\theta)}{d\Omega} = \sum_{ZNJ\pi i} \sigma_{ZN}(J\pi i)\, P_f^{ZN}(J\pi i)\, \frac{dw_{ZN}^{J\pi i}(\theta)}{d\Omega} \tag{10}$$

and calculated, in effect, by the same scheme as the total fission cross Sect. 4.

In practice, it is convenient to define the spin orientation of the nucleus in the ground or excited state with spin J not in terms of the elements of the density matrix $\eta^J(M)$, but in terms of irreducible components of this matrix or, in other words, in terms of spin-tensors of orientation $\tau_{Q0}(J)$ (see e.g. [21]):

$$\tau_{Q0}(J) = \sum_{M} C^{JM}_{JMQ0}\, \eta^{J}(M), \tag{11}$$

where C^{Aa}_{BbDd} are Clebsch–Gordan coefficients. For simplicity, the indices Z, N, π, i that completely define nuclear levels are omitted here. Restoring them and using the identity

$$(2J+1)\left|D^{J}_{MK}(\omega)\right|^{2} = \sum_{Q} (2Q+1)\, C^{JM}_{JMQ0}\, C^{JK}_{JKQ0}\, P_{Q}(\cos\theta) \tag{12}$$

we rewrite (8) as follows

$$\frac{dw^{J\pi i}_{ZN}(\theta)}{d\Omega} = \frac{1}{4\pi} \sum_{Q} (2Q+1)\, \tau^{ZN}_{Q0}(J\pi i)\beta^{ZN}_{Q}(J\pi i)\, P_{Q}(\cos\theta), \tag{13}$$

where $P_Q(\cos\theta)$ are Legendre polynomials; we call the quantities

$$\beta^{ZN}_{Q}(J\pi i) = \sum_{K} C^{JK}_{JKQ0}\, \rho^{J\pi i}_{ZN}(K) \tag{14}$$

the anisotropy parameters. Since $\rho^{J\pi i}_{ZN}(K) = \rho^{J\pi i}_{ZN}(-K)$, they are nonzero only for even values of Q. It is also significant that if the distribution $\rho^{J\pi i}_{ZN}(K)$ (or $\eta^{J\pi i}_{ZN}(M)$) is smooth, then the parameters $\beta^{ZN}_{Q}(J\pi i)$ (or spin-tensors $\tau^{ZN}_{Q0}(J\pi i)$) quickly decrease with increasing of Q.

Thus, substituting (13) to (10), we get for the differential fission cross section

$$\frac{d\sigma_{f}(\theta)}{d\Omega} = \frac{1}{4\pi} \sum_{Q=0,2,4,\ldots} \sigma_{fQ}\, P_{Q}(\cos\theta)\,, \tag{15}$$

where

$$\sigma_{fQ} = (2Q+1) \sum_{ZNJ\pi i} \sigma_{ZN}(J\pi i)\, P^{ZN}_{f}(J\pi i)\, \tau^{ZN}_{Q0}(J\pi i)\beta^{ZN}_{Q}(J\pi i), \tag{16}$$

with $\sigma_{f0} = \sigma_f$. According to (3), the differential fission cross section (15), (16), like the total cross section, devides into the sum of $d\sigma^{DPE}_{f}/d\Omega$ and $d\sigma^{C}_{f}/d\Omega$. As for the angular distribution of the fragments (1), it also takes the form of a series over Legendre polynomials:

$$W(\theta) = \frac{1}{4\pi}\left(1 + \sum_{Q=2,4,\ldots} A_{Q}P_{Q}(\cos\theta)\right), \qquad A_{Q} = \frac{\sigma_{fQ}}{\sigma_{f}}\,. \tag{17}$$

The convenience of these formulae is determined by the fact that in practice almost always at least one of the distributions, either over M or over K, is smooth, so that σ_{fQ} and A_Q decrease rapidly with the growth of Q. Accordingly, with rare exceptions, the shape of the angular distribution $W(\theta)$ is mainly determined by the single parameter A_2.

It has already been said that we have studied the angular distributions $W(\theta)$ of fragments in the fission of a number of isotopes by neutrons with energies of 1–200 MeV. A detailed description of measuring technique and methods of data processing is presented in [1–4]. For all isotopes and at all energies, we have found that only the coefficients A_2 and A_4 are significant in the decomposition (17), and, as expected, the latter is always noticeably smaller than the former. With these coefficients, we have calculated the energy-dependent angular anisotropy

$$\frac{W(0^{\circ})}{W(90^{\circ})} = \frac{1 + A_2 + A_4}{1 - A_2/2 + 3A_4/8} \tag{18}$$

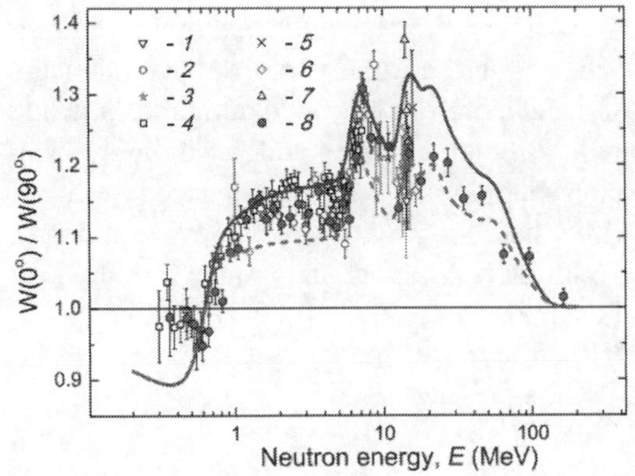

Figure 2. Fission fragment angular anisotropy for ^{237}Np versus neutron energy E. Experimental data: 1 – [22], 2 – [23], 3 – [24], 4 – [25], 5 – [26], 6 – [27], 7 – [28], 8 –[4]. The solid and dashed lines correspond to $\hbar^2/J_{\mathrm{eff}} = 0.022$ and 0.012 MeV.

for each isotope. In Fig. 2 our results are presented on the angular anisotropy of fragments in the reaction ^{237}Np(n, f) together with the experimental values previously obtained for energies up to 20 MeV [22–28]. Note that in some of these works the ratio $W(0°)/W(90°)$ had been measured directly. From Fig. 2 it can be seen that our data are generally in a good agreement with the results of other authors, the data in the region above 20 MeV are obtained for the first time.

The next section is devoted to our calculations of the fission fragment angular distribution (17), i.e. the coefficients A_Q (as expected, they decrease rapidly with the growth of Q). Then, we calculate the angular anisotropy $W(0°)/W(90°)$ and compare it with the observed one. In fact, we thus compare the calculated and observed dominant coefficients A_2, neglecting the contributions to the angular distribution related to $Q = 4$ and above. This is justified at the initial stage of the analysis of angular distributions of fragments. In the following papers we are about to analyze the coefficients A_4.

4 Model for angular distribution of fragments

To calculate the coefficients A_Q (17), (16) one needs to know the spin-tensors of orientation (11) and the anisotropy parameters (14) for all levels of all fissioning nuclei formed in the reaction. As for the spin orientation of the initial compound nuclei, it can be found in the framework of the optical model (see, e.g., [14]; a general expression for the corresponding spin tensors is given in [29]). Transfer of orientation from a decaying nucleus with spin J_1 to a residual nucleus with spin J_2 in the act of emitting of a light particle with a total angular momentum j is described by

$$\tau_{Q0}(J_2) = \sqrt{(2J_1 + 1)(2J_2 + 1)}\; W(LJ_2J_1Q, J_1J_2)\,\tau_{Q0}(J_1), \tag{19}$$

where $W(abcd, ef)$ is Racah function (for the case of gamma transition of multipolarity $L = j$ an expression of this type, presumably, had been first obtained in [30]). Similarly, spin-tensors of orientation of residual nuclei formed at the first stage of the reaction as a result of a direct or pre-equilibrium process can be calculated in the framework of quantum models of these processes.

Note, however, that the calculated cross-section σ_f^C and the observed angular anisotropy of the fission fragments $W(0°)/W(90°)$ decrease similarly at $E > 20$ MeV (see Figs. 1 and 2). Thus, at high energies the angular anisotropy seems be mainly related to the decay of primary compound nucleus. This is so indeed, first, due to the fact that at high collision energy and, consequently, high orbital momentum of incident neutron, the primary compound nucleus possesses a large and well aligned spin J. Second, there is a difference in angular momenta carrying away by particles emitted in statistical decay, on the one hand, and direct and pre-equilibrium processes, on the other hand. In the former case the particles have relatively low energies, of the scale of nuclear temperature, and therefore low angular momenta, while in the later one the particles are usually emitted with high energies and, accordingly, with large angular momenta. Thus, the spin orientation should remain well pronounced for residual nucleus resulting from statistical decay, while there are no grounds for expectations of noticeable spin orientation for residual nuclei resulting from direct and pre-equilibrium processes. Therefore, we take as an assuption that the DPE component of the differential fission cross section is purely isotropic. Thus, instead of (15), (16) we get:

$$\frac{d\sigma_f(\theta)}{d\Omega} = \frac{\sigma_f}{4\pi} + \frac{1}{4\pi} \sum_{Q=2,4,\ldots} \sigma^C_{fQ}\, P_Q(\cos\theta)\,, \tag{20}$$

where the quantities

$$\sigma^C_{fQ} = (2Q+1) \sum_{ZNJ\pi i} \sigma^C_{ZN}(J\pi i)\, P^{ZN}_f(J\pi i)\, \tau^{CZN}_{Q0}(J\pi i)\, \beta^{ZN}_Q(J\pi i) \tag{21}$$

are determined by the population cross sections $\sigma^C_{ZN}(J\pi i)$ and spin-tensors of orientation $\tau^{CZN}_{Q0}(J\pi i)$ for states of either primary compound nucleus or residual nuclei arising at its statistical decay.

Besides the spin-tensors of orientation, the anisotropy parameters $\beta^{ZN}_Q(J\pi)$ (14) are to be calculated. As the starting point, we restrict ourselves by introducing a minimum number of additional parameters, and take the distributions $\rho^{J\pi i}_{ZN}(K)$ to be the same for all Z, N, J, π, and i. Above the barrier we use the statistical distributions, $\rho^{J\pi i}_{ZN}(K) \sim e^{-K^2/2K_0^2}$, where $K_0^2 = J_{\rm eff}T/\hbar^2$ is determined by the temperature T of the nucleus on the barrier and an effective moment of inertia $J_{\rm eff}$ (see [14]), while below the barrier we assume $\rho^{J\pi i}_{ZN}(K) \sim e^{-\alpha(|K|-K_1)^2}$, where α is a fixed parameter and K_1 is the spin projection for the dominant transition state. Note that at high energies of incident neutrons, the excitation energy of a fissioning nucleus, as a rule, sizably exceed the barrier height, and therefore, generally, the type of distribution over K for energies below the barrier is not of importance. Only at low neutron energies, at the sub-barrier fission of the compound nucleus ^{238}Np, the value of K_1 for this isotope essentially determines the type of angular distribution of the fragments; we discuss this point in details below. The choice of K_1 for the isotope ^{237}Np can noticeably affect the angular anisotropy of the fragments in the threshold energy region for the reaction $(n, n'f)$, near $E \simeq 7$ MeV. The experimental data reveal a pronounced peak (see Fig. 2), so to maximize the effect the lowest value $K_1 = 0.5$ is taken. For all other nuclei the value for K_1 was assumed to be 1.5. Additional explanations are presented in [4].

Thus, the quantities (21) for $Q = 2, 4, \ldots$ can be calculated by the same scheme as the compound component of the total fission cross section (5). It was realized as an addition to the TALYS program.

5 Results and discussion

The model described above and the results of first calculations for fission fragment angular distribution in the reaction ^{237}Np(n, f) were previously presented by us in the paper [4]. The main result was that even in its simplest form, with the use of a minimum number of additional parameters, the model successfully reproduces the gross structure of the dependence of the angular anisotropy of fragments on energy in a wide energy range of 1–200 MeV. In this paper, we present some additional results, as well as consider in more details relevant problems.

The first is that at this stage we, in fact, neglect the influence of direct and pre-equilibrium processes on angular anisotropy. There are reasons for this, as shown above, but it would be interesting to obtain quantitative estimates. Note that in [4] we used a simplification that reduces the role of these processes and hence the uncertainties associated with them, namely: the angular distributions of the fragments were computed in neglecting the multiple pre-equilibrium emission. However, as seen from Fig. 1, the account for multiple pre-equilibrium emission only slightly lowers the fission cross section in the region of 20–100 MeV and practically does not affect the compound component σ_f^C of this cross section. Similarly, it turn out that the angular distributions calculated in the framework of the model are scarcely dependent on whether the multiple pre-equilibrium emission is taken into account or not. In this work, the angular distributions are calculated with account for the multiple pre-equilibrium emission.

It has already been mentioned above that in the current simplified version of our model one value is used for the effective moment of inertia on the barrier J_{eff} for all fissioning isotopes. To establish the sensitivity to this parameter, the fission fragment angular anisotropy was calculated for different values of J_{eff}. The results are presented in Fig. 2 for two values with $\hbar^2/J_{\text{eff}} = 0.012$ and 0.022 MeV. These values have been chosen so that to include almost all experimental points above ~1 MeV into region limited by the corresponding curves. As expected, the anisotropy does not depend on J_{eff} at energies below 1 MeV, where the subbarrier fission occurs (in these calculations it is taken $K_1 = 0$ for ^{238}Np), and above 100 MeV, where the observed angular distribution becomes close to isotropic. In reallity, the effective moment of inertia varies from nucleus to nucleus, and this calculation shows the scale of its variation. We plan to include in the model the possibility of taking into account the dependence of J_{eff} on the type of fissioning nuclei. In general, the higher the collision energy, the greater the number of fissioning isotopes contributing to the observed fission cross sections (earlier, the importance of proper account for multi-chance fission was pointed out, in particular, in [12]). Thus, the angular anisotropy of fragments seems to be sensitive to systematics describing not only the effective moment of inertia, but also other parameters significantly dependent on (Z, N), such as the parameters of fission barriers and level density.

Note that in the region 2–16 MeV there are no anomalies in the energy dependence of the calculated angular anisotropy, in contrast to the calculated fission cross section (see Fig. 1). This justifies our assumption about the low sensitivity of the angular distribution to the density of transition states on the barriers.

Special attention is to be paid to the discrepancy between the calculated and observed fission cross-section in the region above 100 MeV – see Fig. 1. It seems likely the models used in TALYS overestimate the role of pre-equilibrium processes in this area, and, thus, underestimate the yield of fissioning primary and secondary compound nuclei. In particular, our experimental data (see Fig. 2) show the presence of a small but nonzero anisotropy above 100 MeV, although the calculated values of $W(0°)/W(90°)$ in this region practically coincide with the unit. Therefore, there is good reason to believe that further studies of the fission fragment angular anisotropy at intermediate energies will refine the contributions of pre-equilibrium processes to cross sections of different processes, in particular, of fission.

Finally, let us turn to energies of incident neutrons below 1 MeV. In this region the fission cross-section falls rapidly due to transition from above-barrier to sub-barrier fission. Here one would expect a significant effect of a low-lying transition state with a definite value of K_1 on the angular distribution of the fragments. However, the reality seems more complicated as follows from the calculations of angular anisotropy performed for different values of K_1 – see Fig. 3 (note that all the curves at this figure have been obtained with $\hbar^2/J_{\text{eff}} = 0.017$ MeV). In the region under consideration, especially below 0.5 MeV, neutron *s*- and *p*-waves dominate. Thus, since the nucleus ^{237}Np has spin and parity $5/2^+$, the compound states 2^+, 3^+, 1^-, 2^-,

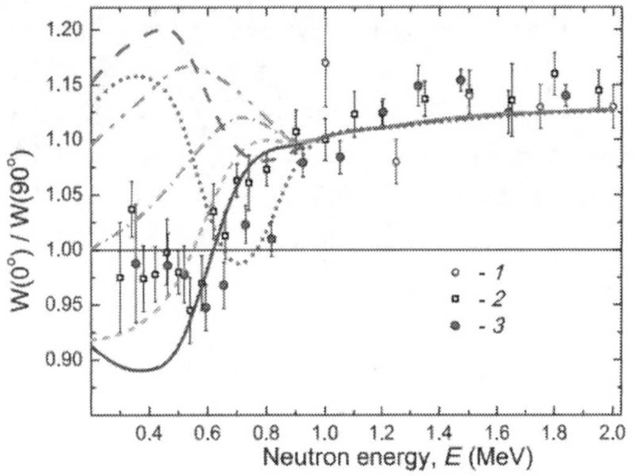

Figure 3. Fission fragment angular anisotropy for ^{237}Np versus neutron energy E for different values of the parameter K_1 of the model (see text). Experimental data: 1 – [23], 2 – [25], 3 –[4]. The solid line corresponds to $K_1 = 0.0$, the short-dashsed line – to $K_1 = 1.0$, the dot-dashed line – to $K_1 = 1.5$, the dot-dot-dashsed line – to $K_1 = 2.0$, the long-dashed line – to $K_1 = 3.0$, the dotted line – to $K_1 = 4.0$.

3^-, 4^- are predominantly excited in neutron capture. So the spin projection K_1 was taken in the range from 0 to 4. Although the accuracy of the experimental data is not perfect, it can be seen that none of the curves describes them satisfactorily. The most appropriate are the extreme values K_1, equal to 0 and 4. Thus, in the sub-barrier energy region our approach to distribution over K does not look suitable. However, do note that the data [25] had been obtained a long time ago, but no successful attempt to interpret them is known.

6 Summary

Angular distributions of fragments in neutron-induced nuclear fission have attracted the attention of researchers for many decades. Substantial progress was achieved in [12], where for the first time not only experimental data for energies noticeably exceeding 20 MeV were obtained, but also an attempt was made to describe the data using modern computer codes on the basis of a model similar to that described in this paper. This model, in particular, was based on the calculations of the population cross sections of nuclear states with the spin projections M on the z axis. The results obtained for the reactions ^{232}Th (n, f) and ^{238}U (n, f) were compared with the experimental data presented in the same paper. However, no further publications of these authors came out, to unveil the necessary details on the method used or perform calculations for other target nuclei.

In recent years, invetigations have been extended to intermediate energies, up to 200 MeV, as well as to a wide range of target isotopes. The need to systematize new data, extract from them essential information about the fission process as well as the features of neutron-nucleus reactions at intermediate energies prompted us to develope a method for calculating the fission fragment angular distributions. In this paper we present the method based on the TALYS program and show that the results of calculations based on the simplified model are in good agreement with experimental data. It provides a hope that new information about fission barriers, transition states, as well as the role of pre-equilibrium processes in the neutron-nuclear interaction can be obtained by improving and complicating the model.

At this stage, we see the need to complicate our model by including an account for the dependence of parameters such as the effective moment of inertia on the type of fissioning nuclei. It would be useful to develop methods for calculating the spin-tensors of orientation for excited nuclei formed in direct and pre-equilibrium processes. At low energy region, it is necessary to combine in some way an evaluation of distribution of fission probabilities over K with the usually performed calculation of fission transmission coefficients involving account for appropriate transition states. The contribution related to the fourth Legendre polynomial into fission fragment angular distribution calls for further investigation. In general, it seems that nowadays the experimental data on angular distributions of fragments are little used for improving the current fission model, thus the main problem is to correct this situation.

At last, mention that the developed methods can be used to calculate the angular distributions of γ-quanta emitted by excited compound nuclei, which, for example, may be useful in the study of the $(n, \gamma f)$ reaction. In addition, all these methods can be extended to reactions initiated by charged particles, such as (p, f) and (p, γ).

Partial support from the RFBR Grant No. 18-02-00571 is greatly acknowledged.

References

[1] A.S. Vorobyev, A.M. Gagarski, O.A. Shcherbakov, L.A. Vaishnene, A.L. Barabanov, JETP Lett. **102**, 203 (2015)

[2] A.S. Vorobyev, A.M. Gagarski, O.A. Shcherbakov, L.A. Vaishnene, A.L. Barabanov, JETP Lett. **104**, 365 (2016)

[3] A.S. Vorobyev, A.M. Gagarski, O.A. Shcherbakov, L.A. Vaishnene, A.L. Barabanov, JETP Lett. **107**, 521 (2018)

[4] A.S. Vorobyev, A.M. Gagarski, O.A. Shcherbakov, L.A. Vaishnene, A.L. Barabanov, JETP Lett. **110**, 242 (2019)

[5] A.S. Vorobyev, A.M. Gagarski, O.A. Shcherbakov, L.A. Vaishnene, A.L. Barabanov, EPJ Web Conf. **146**, 04011 (2017)

[6] A.S. Vorobyev, A.M. Gagarski, O.A. Shcherbakov, L.A. Vaishnene, A.L. Barabanov, Bull. RAS Phys. **82**, 1240 (2018)

[7] N.K. Abrosimov, G.Z. Borukhovich, A.B. Laptev, V.V. Marchenkov, G.A. Petrov, O.A. Shcherbakov, Yu.V. Tuboltsev, V.I. Yurchenko, Nucl. Instrum. Methods Phys. Res. A **242**, 121 (1985)

[8] O.A. Shcherbakov, A.S. Vorobyev, E.M. Ivanov, Phys. Part. Nucl. **49**, 81 (2018)

[9] D. Tarrio, L.S. Leong, L. Audouin, I. Duran, C. Paradela *et al.*, Nucl. Data Sheets **119**, 35 (2014)

[10] E. Leal-Cidoncha, I. Duran, C. Paradela, D. Tarrio, L.S. Leong *et al.*, EPJ Web Conf. **111**, 10002 (2016)

[11] V. Geppert-Kleinrath, F. Tovesson, J.S. Barrett, N.S. Bowden, J. Bundgaard *et al.*, Phys. Rev. C **99**, 064619 (2019)

[12] I.V. Ryzhov, M.S. Onegin, G.A. Tutin, J. Blomgren, N. Olsson, A.V. Prokofiev, P.-U. Renberg, Nucl. Phys. A **760**, 19 (2005)

[13] A. Bohr, *Proc. Int. Conf. on the Peaceful Uses of Atomic Energy, Geneva, 1955, v. 2* (UN, New York, 1956) p. 151

[14] R. Vandenbosch, J.R. Huizenga, *Nuclear Fission* (Academic Press, New York, 1973) Chapters V and VI

[15] A.J. Koning, S. Hilaire, M.C. Duijvestijn, *Proc. Int. Conf. on Nuclear Data for Science and Technology, Nice, 2007* (EDP Sciences, 2008), p. 211

[16] R. Capote, M. Herman, P. Oblozinsky, P.G. Young , S. Goriely *et al.*, Nucl. Data Sheets **110**, 3107 (2009)

[17] O. Shcherbakov, A. Donets, A. Evdokimov, A. Fomichev, T. Fukahori *et al.*, J. Nucl. Sci. Tech. Suppl. **39**, 230 (2002)

[18] C. Paradela, L. Tassan-Got, L. Audouin, B. Berthier, I. Duran *et al.*, Phys. Rev. C **82**, 034601 (2010)

[19] M. Diakaki, D. Karadimos, R. Vlastou, M. Kokkoris, P. Demetriou *et al.*, Phys. Rev. C **93**, 034614 (2016)

[20] H.J. Krappe, K. Pomorski, *Theory of Nuclear Fission* (Springer, Heidelberg, 2012) Chapter 4

[21] K. Blum, *Density matrix theory and applications* (Springer, Heidelberg, 2012), Chapter 4

[22] J.E. Brolley, Jr., W.C. Dickinson, Phys. Rev. **94**, 640 (1954)

[23] J.E. Simmons, R.L. Henkel, Phys. Rev. **120**, 198 (1960)

[24] R.B. Leachman, L. Blumberg, Phys. Rev. **137**, B814 (1965)

[25] D.L. Shpak, B.I. Fursov, G.N. Smirenkin, Sov. J. Nucl. Phys. **12**, 19 (1971)

[26] R.H. Iyer, M.L. Sagu, *Proc. Nucl. and Solid State Physics Symp., Madurai, 1970, v. 2* (1970), p. 57

[27] Kh.D. Androsenko, G.G. Korolev, D.L. Shpak, VANT Ser. Yadernye Konstanty **46**, 9 (1982)

[28] S. Ouichaoui, S. Juhasz, M. Varnagy, J. Csikai, Acta Phys. Hungarica **64**, 209 (1988)

[29] A.L. Barabanov, D.P. Grechukhin, Sov. J. Nucl. Phys. **43**, 892 (1986)

[30] J.A.M.Cox, H.A.Tolhoek, Physica **19**, 673 (1953)

Multiplicity of scision neutrons from density functional scission dynamics

Nicolae Carjan[1,2,*], *Ionel* Stetcu[3], *Margarit* Rizea[2], and *Aurel* Bulgac[4]

[1]Joint Institute of Nuclear Research, Dubna, Russia
[2]"Horia Hulubei" National Institute for Physics and Nuclear Engineering, Bucharest, Romania
[3]Theoretical Division, Los Alamos National Laboratory, Los Alamos, NM 87545, USA
[4]Department of Physics, University of Washington, Seattle, Washington 98195, USA

Abstract. The time evolution of the nuclear density of the fissioning system ^{240}Pu during the scission process is obtained from the time-dependent superfluid local-density approximation (TDSLDA) to the density functional theory. A nuclear energy density functional based on the Skyrme force Skm* is used. The duration of the scission process Δt as well as the neck radius ($r_{\rm min}$) of the 'just-before scission' configuration and the minimum separation ($d_{\rm min}$) of the inner surfaces of the fragments in the 'immediately-after scission' configuration were extracted in order to calculate the multiplicity of the scission neutrons ($\nu_{\rm sc}$) using a phenomenological dynamical scission model (DSM). We find that $\nu_{\rm sc}$=1.347, i.e. half of the prompt fission neutrons measured in the reaction ^{239}Pu($n_{\rm th}$, f) are released at scission. After scission, the fragments are left excited and with some extra deformation energy (mainly the heavy one). In this way we can account for the evaporation of the other half and for the emission of γ rays.

1 Introduction

The most important feature of nuclear fission is that it is accompanied by emission of prompt fission neutrons (PFN). Two PFN sources have been discussed so far:

1) Neutrons evaporated from fully-accelerated excited fragments (EVN).

2) Neutrons dynamically released at scission (SN) due to the diabatic coupling between the neutron degree of freedom and the rapidly changing neutron-nucleus potential.

The relative intensities of these two components are unknown. It is very difficult to separate them experimentally since the gross features of PFN can be accounted for by both hypotheses.

While the 1st component depends on the excitation and extra deformation with which the fission fragments are born [1], the 2nd depends on the dynamical evolution of the system during the scission process [2]. Therefore a simple stationary approach suffices for EVN but a more complicated dynamical approach is needed for SN.

There are three parameters in the dynamical scission model: the nuclear shapes just before (α_i) and immediatelly after (α_f) scission and the duration ΔT of the transition between these two shapes. Since these quantities were unknown, in the past we have used educated guesses.

*e-mail: carjan@theory.nipne.ro

At present reliable values for these quantities are available from the density functional theory (DFT) applicable to superfluid systems (usually named the superfluid local density approximation or SLDA), within a time-dependent extension (TD-SLDA) [3]. Among the many nuclear energy density functionals we choose the one based on the Skyrme force SkM* [4] that reproduces the fission barriers in ^{240}Pu. We extract these three quantities and calculate the multiplicity of the neutrons released at scission and emitted immediately after (i.e., during the acceleration of the fragments).

Finally we also estimate the energy available to emit prompt γ rays and evaporate neutrons: $E^*_{\mathrm{sc}} + \Delta E_{\mathrm{def}}$

2 Dynamical scission model (DSM)

To simulate the evolution of the neutron wave functions from α_i to α_f lasting ΔT, the two-dimensional time-dependent Schrödinger equation (TDSE2D) with a time-dependent potential (TDP) is solved. The distribution of final wave packets $\hat{\Psi}^i(\Delta T)$ (that correspond at $t = 0$ to the eigenstates $\hat{\Psi}^i$ of α_i) over the eigenstates $\hat{\Psi}^f$ of α_f:

$$a_{if} = \langle \hat{\Psi}^i(\Delta T) | \hat{\Psi}^f \rangle. \tag{1}$$

constitutes the core of DSM. All wave functions have an implicit dependence on the cylindrical coordinates (ρ, z). a_{if} is $\neq 0$ only if $|\hat{\Psi}^i\rangle$ and $|\hat{\Psi}^f\rangle$ have the same projection Ω of the total angular momentum along the symmetry axis.

The rapid change of the potential in which the nucleons move produces, by means of a diabatic coupling, their excitation and eventual emission.

The a_{if} coefficients are used to calculate the scission neutron multiplicity ν_{sc} in a simple and intuitive way. It is given by the sum of the probabilities P^i_{em} that a neutron occupying a given bound-state i is emitted:

$$P^i_{\mathrm{em}} = v_i^2 \textstyle\sum_f (1 - |a_{if}|^2)$$

$$\nu_{\mathrm{sc}} = 2 \sum_i P^i_{\mathrm{em}} \tag{2}$$

The sums are over the bound states of the initial and final configurations. v_i^2 is the occupation probability of the state i.

Similarly, the excitation energy in which the fragments are left immediately after scission is given by

$$E^*_{\mathrm{sc}} = 2 \times (\sum_f V_f^2 e_f - \sum_f v_f^2 e_f), \tag{3}$$

where $V_f^2 = \sum_i v_i^2 |a_{if}|^2$ is the probability that an eigenstate $|\hat{\Psi}^f\rangle$ of the final configuration α_f is occupied after the sudden transition and e_f is its eigen-energy. The i and f sums are over the bound states of the initial and final nuclear configurations, respectively. The nuclear shapes involved are described by Cassini ovals [5] with two parameters: α (the overall deformation) and α_1 that fixes the mass asymmetry.

3 TD-SLDA

In SLDA, the nuclear system is described by a density functional $\mathcal{E}(n(\vec{r}), \nu(\vec{r}), \vec{j}(\vec{r}), \ldots)$, where n is the normal (number) density, ν is the anomalous density, $\vec{j}$ is the current density, etc. They are defined in terms of the $u_{k\uparrow(\downarrow)}$ and $v_{k\uparrow(\downarrow)}$ components of the quasiparticle wave functions (qpwfs). The dynamics of the nucleus is followed in real time by solving a system of time-dependent Schrödinger-like equations

$$i\hbar\frac{\partial}{\partial t}\begin{pmatrix} u_{k\uparrow} \\ u_{k\downarrow} \\ v_{k\uparrow} \\ v_{k\downarrow} \end{pmatrix} = \tag{4}$$

$$\begin{pmatrix} h_{\uparrow\uparrow}-\mu & h_{\uparrow\downarrow} & 0 & \Delta \\ h_{\downarrow\uparrow} & h_{\downarrow\downarrow}-\mu & -\Delta & 0 \\ 0 & -\Delta^* & -(h^*_{\uparrow\uparrow}-\mu) & -h^*_{\uparrow\downarrow} \\ \Delta^* & 0 & -h^*_{\downarrow\uparrow} & -(h^*_{\downarrow\downarrow}-\mu) \end{pmatrix}\begin{pmatrix} u_{k\uparrow} \\ u_{k\downarrow} \\ v_{k\uparrow} \\ v_{k\downarrow} \end{pmatrix},$$

with the dependence on the spacial coordinates and time not shown explicitly. The one-body Hamiltonian $h = \partial\mathcal{E}/\partial v^*$ and the pairing field Δ depend on the densi- ties and currents constructed from the qpwfs at the time t.

3.1 TD-SLDA scission dynamics

We calculate at each time the number density
$n(\vec{r},t) = \sum_k(|v_{k\uparrow}(\vec{r},t)|^2 + |v_{k\downarrow}(\vec{r},t)|^2)$ shown in Fig. 1 at several times before and after scission for ^{240}Pu. This sequence of densities leads to a mass asymmetry of the separated fragments defined by A_L=104 (A_H=136) close to the most probable experimental mass division in the reaction ^{239}Pu(n_{th}, f). The middle equidensity line is the half density $n_0/2$ with n_0=0.16 fm^{-3}. It is equivalent with the equipotential line $V_0/2$ of the deformed Woods-Saxon potential used in DSM. V_0 is the potential depth. These equipotential lines are shown for comparison on the same figure. One can notice a good similarity between the nuclear shapes used in the two models.

3.2 The limits of the scission process

To determine when the scission process starts and when it ends we analyze the evolution of the moments of the density distribution, in particular we look for the sharp decrease of the total quadrupole and for the oscillation of the fragment octupoles (see Fig. 2).

4 Parametrization of the scission process

One can estimate that the rapid transition at scission starts at ≈800 fm/c and lasts untill ≈1200 fm/c, its duration being $\Delta T = 13.4 \times 10^{-22}$ sec. Going back to Fig. 1, one can see that the corresponding configurations have $r_{\min}$ = 2.13 fm and $d_{\min}$ = 11.48 fm respectively. $r_{\min}$ and $d_{\min}$ are the neck radius and the separation of the inner surfaces of the fragments respectively.

This inference is also supported by the variation of the octupole moments Q_{zzz} of the L and H fragments from Fig. 2.

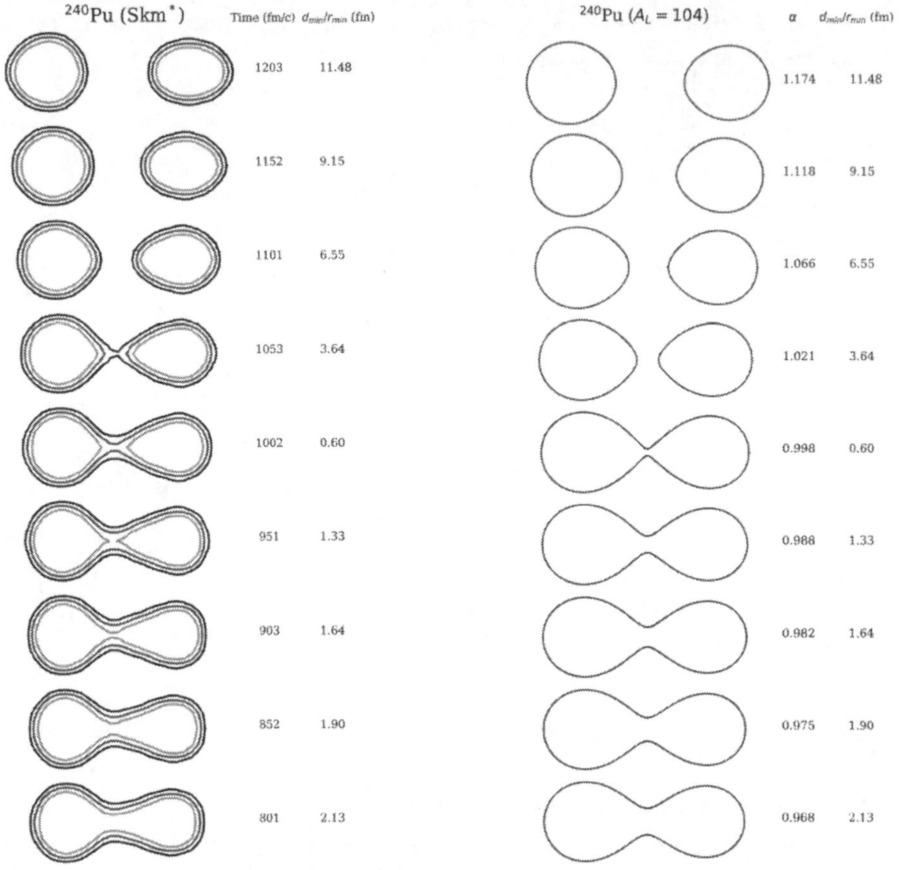

Figure 1. Time evolution of the density (left column) and of the potential (right column).

Table 1. Scission neutron multiplicity:$^{240}Pu, A_L$ = 104, 0.968 >> 1.174, ΔT = 13.4 × 10^{-22} sec.

Ω	1/2	3/2	5/2	7/2	9/2	11/2	$\Sigma(\Omega)$
ν_{sc}	0.833	0.465	0.037	0.005	0.006	0.001	1.347

4.1 SCN multiplicities

The deformations of the Cassini ovals that have the same $r_{\min}$ and $d_{\min}$ values are found (Fig. 1) to be $\alpha_i = 0.968$ and $\alpha_f = 1.174$ respectively. We therefore calculate the evolution of all neutron wave functions from α_i to α_f keeping the mass asymmetry fixed at A_L=104 with the help of the parameter α_1. The duration of the transition is ΔT=13.4 10^{-22} sec.

4.2 SCN, EVN and γ rays

The resulting scission-neutron multiplicities are presented for each set of wave functions of given Ω in Table 1. ν_{sc} decreases with Ω, the contribution of Ω = 1/2 being more then 50%. In the case of ^{240}Pu the multiplicity summed over all Ω values is found to be 1.347, i.e., half of the prompt fission neutrons measured in the reaction ^{239}Pu(n_{th}, f) are released at scission.

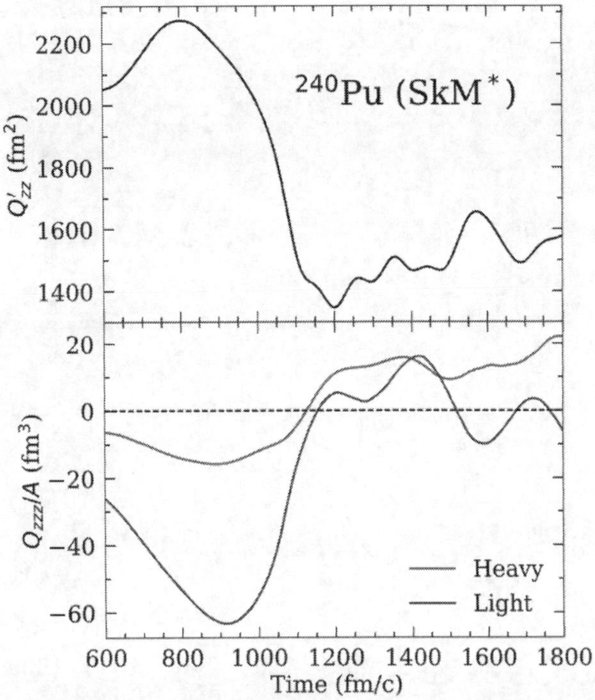

Figure 2. Evolution of the quadrupole moment (of the whole system) and of the octupole moment (of the fragments) from before to after scission.

Table 2. Deformation parameters of the primary fragments and their extra deformation energy: $^{240}Pu, A_L = 104$.

Z	A	β_2	β_3	ΔE_{def}
41	104	0.39	0.10	0.79 MeV
53	136	0.10	0.10	2.27 MeV

In addition of being promoted to positive energy states at the end of the scission process, the neutrons are also promoted to states between the Fermi and the "zero" level producing excitation in the primary fragments. Eq. (3) gives $E^*_{\mathrm{sc}} = 12.59$ MeV.

Finally we estimate the extra deformation energy ΔE_{def} of the primary fragments. From the figures with the time evolution of Q_{zz} and Q_{zzz} of each fragment, one can extract their deformations β_2 and β_3 (see Fig. 3) and from this we calculate their post-scission deformation energies [6]. The results are summarized in Table 2.

On one side we have the estimated total excitation energy

$E^*_{\mathrm{sc}} + \Delta E_{\mathrm{def}} = 15.65$ MeV

that will be used to emit prompt γ rays and evaporate neutrons.

On the other side we have $< E_\gamma > = 6.70$ MeV measured [7] and 1.53 neutrons to evaporate. We use experimental masses to calculate neutron separation energies. $S_n = 4.97$ MeV and 3.78 MeV for the L and H fragments for the odd-odd L and H fragments from Table 2.

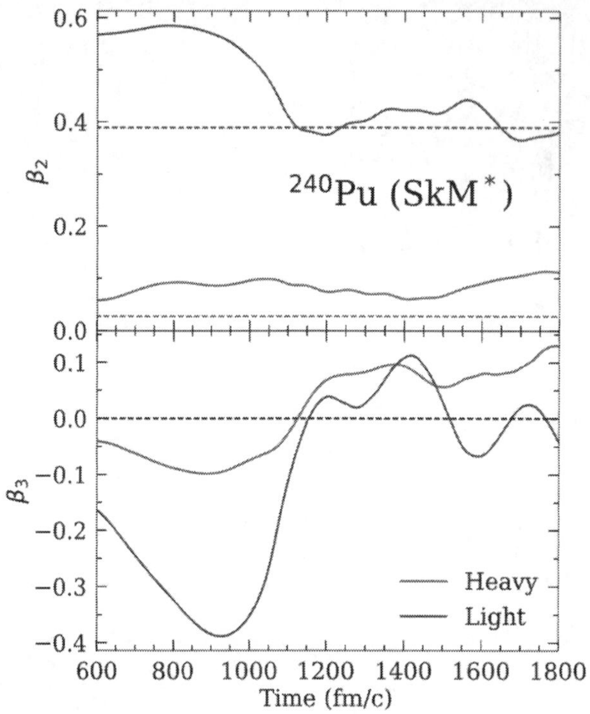

Figure 3. Evolution of β_2 and β_3 deformations of the nascent fragments.

For neighbouring even-even fragments S_n = 7.55 MeV and 4.67 MeV. Using an average value for S_n, the evaporation of 1.53 neutrons requires 8.02 MeV. So the sum is 14.72 MeV.

In view if the uncertainties in the model one can consider the agreement between the total excitation energy calculated (15.65 MeV) and the energy carried by evaporated neutrons and gammas (14.72 MeV) satisfactory.

5 Discussion and conclusions

• We give a first plausible answer to a long-standing question regarding the emission of prompt fission neutrons: is the percentage of neutrons emitted at scission and during acceleration significant? This answer is YES!

• We use DSM to calculate ν_{sc} and E^*_{sc} In this phenomenological model the dynamics of the neck rupture determines how many unbound neutrons and how much excitation we have at the end.

• Implementing the scission dynamics given by TD-SLDA into DSM, we obtain ν_{sc}= 1.347 (half of the PFN originate at scission) and E^*_{sc} = 12.59 MeV.

• From the deformations of the primary fragments predicted by TD-SLDA, we calculated their extra deformation energies ΔE_{def} = 3.06 MeV.

• The total energy balance is fulfilled to a good approximation.

We are grateful to Patrick Talou for his interest in this work and very useful discussions.

References

[1] D.G. Madland, J.R. Nix, Nucl. Sci. Eng. **81**, 213 (1982)

[2] N. Carjan, M. Rizea, Phys. Rev. C **82**, 014617 (2010)
[3] A. Bulgac, P. Magierski, K.J. Roche, I. Stetcu, Phys. Rev. Lett. **116**, 122504 (2016)
[4] J. Bartel, P. Quentin, M. Brack, C. Guet, H.-B. Håkansson, Nucl. Phys. A **386**, 79-100 (1982)
[5] V.V. Pashkevich, Nucl. Phys. A **169**, 275-293 (1971)
[6] P. Möller, A.J. Sierk, T. Ichikawa, H. Sagawa, Atomic Data and Nuclear Data Tables, **109-110**, 1-204 (2016)
[7] H.H. Knitter, U. Brosa, C. Butz-Jorgensen, *The Nuclear Fission Process* (CRC Press, Boca Raton USA, 1991), p. 497

The study of prompt fission γ rays at the Oslo Cyclotron Laboratory

D. Gjestvang[1,*], S. Siem[1], F. Zeiser[1], J. Randrup[2], R. Vogt[3,4], F. Bello-Garrote[1], L.A. Bernstein[2,5], D.L. Bleuel[3], M. Guttormsen[1], A. Görgen[1], A.C. Larsen[1], K.L. Malatji[6,7], E. Matthews[5], V. Modamio[1], A. Oberstedt[8,9], S. Oberstedt[10], T. Tornyi[11], G. Tveten[1], A. Voyles[5], and J. Wilson[12]

[1]Department of Physics, University of Oslo, Oslo, Norway
[2]Nuclear Science Division, LBNL, Berkeley, USA
[3]Nuclear and Chemical Sciences Division, LLNL, Livermore, USA
[4]Physics Department, University of California, Davis, USA
[5]Nuclear Engineering Department, University of California, Berkeley, USA
[6]iThemba LABS, Somerset West, South Africa
[7]Physics Department, University of Stellenbosch, Matieland, South Africa
[8]ELI-NP/IFIN-HH, Bucharest-Magurele, Romania
[9]Department of Physics and Astronomy, Uppsala University, Sweden
[10]European Commission, Joint Research Center for Nuclear Safety and Security, Geel, Belgium
[11]MTA-ATOMKI, Debrecen, Hungary
[12]IPN Orsay, Orsay Cedex, France

Abstract. The study of prompt fission γ rays (PFGs) is crucial for understanding the energy and angular momentum distribution in fission, and over the last decade there has been an revived interest in this aspect of fission. We present the new experimental setup at the Oslo Cyclotron Laboratory for detecting PFGs resulting from charged particle-induced fission. Additionally, PFGs from the reaction ^{240}Pu(d,pf) were measured in April 2018, and the fission gated proton-γ coincidence spectrum is shown. In order to explore the dependence of the PFG emission on the excitation energy and angular momentum of the compound nucleus, we plan several experiments where charged particle reactions are used to induce fission in various plutonium isotopes. The final results will be compared to predictions made by the Fission Reaction Event Yield Algorithm (FREYA) in an upcoming publication, to benchmark the current modelling of both the PFGs and the fission process.

1 Introduction

In 2019 we celebrate 80 years since the discovery of fission, but still a lot of properties of fission are not quite understood. One of these aspects is the emission of prompt fission γ rays (PFGs) which are photons emitted from the fission fragments as they de-excite. These photons carry valuable information about the fissioning system [1], and by examining them, useful insights can be obtained as we try to understand how fission unfolds. Furthermore,

*e-mail: dorthea.gjestvang@fys.uio.no

a detailed understanding of PFG emission is crucial for reactor applications. The prompt photons, though carrying only a small percentage of the total energy released in fission [2], might deposit energy far from where they are emitted. Thus, good knowledge of the PFGs is essential when designing the next generation of nuclear reactors. Especially the fast neutron region is important, where the PFG characteristics are the least studied [3].

The prompt fission γ rays are characterised by their photon spectrum (PFGS), from which the quantities the average total energy released per fission $E_{\rm tot}$, the average multiplicity per fission $M_{\rm g}$, the average photon energy $E_{\rm g}$, can be extracted. Over the last decade, several measurements of these PFG characteristics have been conducted, see for example Refs. [3–6], and the focus has been on determining PFGs from neutron-induced fission of key actinides. Theoretical efforts have also been made to model PFG emission [7–9]. By comparing model predictions to experimental measurements, we can benchmark the current modelling of photon emission from fission.

In order to get a broader perspective on the behaviour of the PFGs, the Oslo Cyclotron Laboratory (OCL) has started a campaign to measure the PFGs resulting from charged particle-induced fission, which has several advantages compared to neutron-induced fission. The energy of the charged projectile can easily span a range of energies, and by detecting the energy of the ejectile the excitation energy $E_{\rm x}$ of the fissioning nucleus can be determined. Charged particle-induced fission thus enables us to inspect the PFG dependence on $E_{\rm x}$. Furthermore, surrogate reactions with charged particles are expected to induce more angular momentum in the compound nucleus [10]. Thus, by comparing PFG characteristics measured through different surrogate reactions, their dependence on angular momentum can be explored. The OCL experimental setup combines a charged particle detector, fission detectors, and a state-of-the-art $LaBr_3$ scintillator detector array which makes it a powerful tool for studying prompt fission γ rays through charged particle-induced fission.

2 Experimental setup for PFG detection at the OCL

The Oslo Cyclotron Laboratory has an MC-35 Scanditronix cyclotron which can deliver a pulsed beam of protons, deuterons, ^{3}He, or α particles. The charged particle beam is steered to the experimental hall, where it induces various reactions in the target. When studying prompt fission γ rays, fission events are selected using fission fragment detectors that are PPACs (Parallel Plate Avalanche Counter). These are gas-filled, low pressure detectors sensitive to heavy ions only [11], and are placed 5 cm in the forward direction, see figure 1. The placement of the PPACs is important, as the fission fragments are easily stopped and cannot travel through the supporting backing of the targets. For the same reason, the reaction chamber must be in vacuum. The ejectile from the charged particle reaction is detected by the silicon detector ring SiRi, placed in the backward position. SiRi consists of 64 particle telescopes with a front-thickness of 130 μm and back-thickness of 1550 μm [12], and by observing the energy and angle of the ejectile, the compound nucleus excitation energy is reconstructed.

The prompt fission γ rays are detected by the Oslo SCintillator ARray (OSCAR), consisting of 30 inorganic cerium-doped $LaBr_3$ scintillator detectors. The detectors are cylinders, and each measures 88.9 mm × 203 mm (diameter × length). The energy resolution of these detectors is about twice as good as the resolution for NaI(Tl) detectors, with $< 2.5\%$ resolution for 1.3 MeV γ rays. The decay time is also about one-tenth of the decay time of NaI(Tl). As the previous photon detector array (CACTUS) at the OCL consisted of NaI(Tl) detectors, this is a major improvement in both the time and energy resolutions, and opens for new experiments to be performed. In figure 1 the layout of the experimental setup is illustrated.

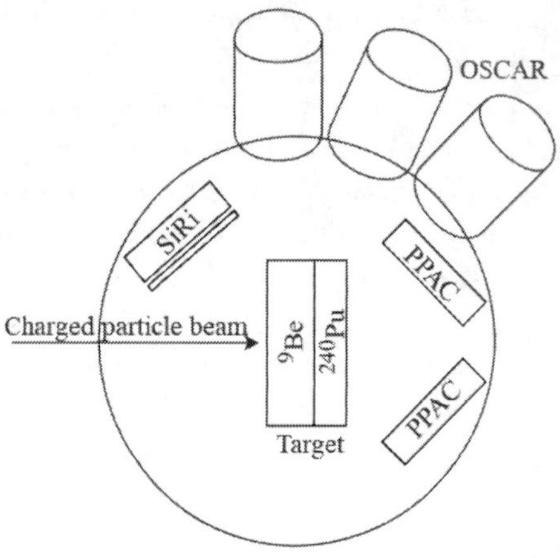

Figure 1. Experimental setup for detecting PFGs at the OCL, illustrated with the target used in the April 2018 experiment. Three of the 30 $LaBr_3$ detectors forming OSCAR are shown.

2.1 The ^{240}Pu(d,pf) experiment

The first PFG experiment using OSCAR was run in April 2018. A target of 0.4 mg/cm^2 thick ^{240}Pu, placed on a backing of 2.3 mg/cm^2 ^{9}Be, was bombarded with 13.5 MeV deuterons and the resulting PFGs were measured. The distance between OSCAR and the target was chosen to be 20 cm for most of the detectors. The majority of the prompt fission neutrons, which can create photon-like signals in the $LaBr_3$ detectors, could be rejected through the time-of-flight technique. One detector was pulled back to 40 cm to further test the ability to discriminate between neutrons and γ rays. At this time, 28 of the 30 $LaBr_3$ detectors were installed.

3 Results

The data analysis for the ^{240}Pu(d,pf) experiment is ongoing and will be published in an upcoming paper. Figure 2 shows the raw data matrix, with the number of γ rays in coincidence with both a fission fragment and a proton plotted as a function of compound nucleus excitation energy E_x and measured photon energy E_γ. Note that E_γ represents the measured energy of a photon and is different from E_g which is the average PFG energy. At the conference, the preliminary PFG photon spectrum was shown, plotted for different ^{241}Pu excitation energy bins. The photon spectrum was compared to predictions made with FREYA (Fission Reaction Event Yield Algorithm) which provides a complete description of fission where all physical quantities are conserved [13]. Preliminary extracted values for E_{tot}, M_g, and E_g were shown, plotted as a function of E_x.

In order to obtain the true photon spectrum of the PFGs, the measured spectrum must be unfolded using the detector response function. The detector response function of OSCAR is not yet properly determined for photon energies below 500 keV. This is unfortunate, as the low-energy photons have been shown to have a large effect on the determined PFG characteristics [14]. To correct this, measurement of the PFGS from the spontaneous fission of ^{252}Cf is planned. As the PFGS from ^{252}Cf(sf) is well known, it acts as a benchmark for the experimental setup and the data analysis routine.

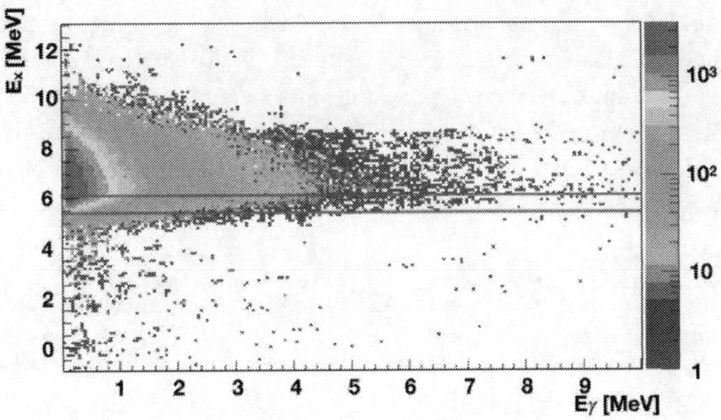

Figure 2. The coincidence matrix, showing the energies E_γ of γ rays emitted from ^{240}Pu(d,pf) events on the x-axis, plotted against the excitation energy E_x of ^{241}Pu* on the y-axis. The colors on the right-hand side of the plot show the number of photons in each (E_γ,E_x) bin. The red lines represent the values for the double-humped fission barrier, as reported in Ref. [15, pg. 800].

4 Outlook

A range of PFG experiments are scheduled to run at the OCL with the new experimental setup that includes OSCAR. In the spring of 2020, PFG characteristics will be extracted from the ^{239}Pu(d,pf) reaction. The PFGs resulting from ^{239}Pu(n,f) have been measured multiple times [4, 16–18], and (d,pf) is a surrogate reaction for (n,f). Even though both reactions produce the compound nucleus ^{240}Pu*, (d,pf) is expected to impart more angular momentum to the fissioning nucleus, allowing us to study the unknown impact of additional angular momentum on the PFG characteristics.

^{239}Pu(d,pf) has previously been run at the OCL and results are published by Rose *et al.* in Ref. [10]. However, the old photon detector array CACTUS was used. The time resolution of these NaI detectors did not allow for a separation between prompt neutrons and photons through TOF, and the photon detection threshold was 400 keV. Therefore, we repeat the experiment to both obtain PFG characteristics below 400 keV and to improve the experimental results above this threshold.

Another experiment, scheduled for spring 2020, is to induce fission in a ^{244}Pu target through a variety of charged particle reactions. This allows for the production of exotic actinides and studying their PFGs will further broaden our understanding of photon emission from the fission fragments.

Thus, we plan to study prompt fission γ rays over a wide range of compound nucleus excitation energies and angular momenta. Through new experiments, and comparisons between experimental results and FREYA predictions, we expect to gain new insights into the behaviour of prompt fission γ rays and therefore also the fission process.

The authors would like to thank J. C. Möller, P. Sorbas, and J. C. Wikne at the Oslo Cyclotron Laboratory for providing the high-quality deuteron beam during the experiment. This work was supported by the Research Council of Norway under project Grant No. 263030, and under the auspices of the U.S. Department of Energy by Lawrence Livermore National Laboratory under Contract DE-AC52-

07NA27344. A. C. L. gratefully acknowledges funding through ERC-STG-2014 under grant agreement no. 637686. The work of R.V. was performed under the auspices of the U.S. Department of Energy by Lawrence Livermore National Laboratory under Contract DE-AC52-07NA27344. The work of J.R. was performed under the auspices of the U.S. Department of Energy by Lawrence Berkeley National Laboratory under Contract DE-AC02-05CH11231. This work was supported by the Office of Defense Nuclear Nonproliferation Research & Development (DNN R&D), National Nuclear Security Administration, U.S. Department of Energy.

References

[1] D. Choudhury, A. Gatera, A. Göök, M. Lebois, A. Oberstedt, S. Oberstedt, L. Qi, J. Wilson, AIP Conf. Proc. **2076**, 060002 (2019)

[2] G. Rimpault, D. Bernard, D. Blanchet, C. Vaglio-Gaudard, S. Ravaux, A. Santamarina, Phys. Procedia **31**, 3 (2012)

[3] L. Qi, M. Lebois, J.N. Wilson, A. Chatillon, S. Courtin, G. Fruet, G. Georgiev, D.G. Jenkins, B. Laurent, L. Le Meur et al., Phys. Rev. C **98**, 014612 (2018)

[4] A. Gatera, T. Belgya, W. Geerts, A. Göök, F.J. Hambsch, M. Lebois, B. Maróti, A. Moens, A. Oberstedt, S. Oberstedt et al., Phys. Rev. C **95**, 1 (2017)

[5] A. Oberstedt, T. Belgya, R. Billnert, R. Borcea, T. Bryš, W. Geerts, A. Göök, F.J. Hambsch, Z. Kis, T. Martinez et al., Phys. Rev. C **87**, 051602(R) (2013)

[6] M. Lebois, J.N. Wilson, P. Halipré, A. Oberstedt, S. Oberstedt, P. Marini, C. Schmitt, S.J. Rose, S. Siem, M. Fallot et al., Phys. Rev. C **92**, 034618 (2015)

[7] R. Vogt, J. Randrup, Phys. Rev. C **96**, 064620 (2017)

[8] I. Stetcu, P. Talou, T. Kawano, M. Jandel, Phys. Rev. C **90**, 024617 (2014)

[9] M.J. Marcath, R.C. Haight, R. Vogt, M. Devlin, P. Talou, I. Stetcu, J. Randrup, P.F. Schuster, S.D. Clarke, S.A. Pozzi, Phys. Rev. C **97**, 044622 (2018)

[10] S.J. Rose, F. Zeiser, J.N. Wilson, A. Oberstedt, S. Oberstedt, S. Siem, G.M. Tveten, L.A. Bernstein, D.L. Bleuel, J.A. Brown et al., Phys. Rev. C **96**, 014601 (2017)

[11] T. Tornyi, A. Görgen, M. Guttormsen, A. Larsen, S. Siem, A. Krasznahorkay, L. Csige, Nucl. Instrum. Meth. A **738**, 6 (2014)

[12] M. Guttormsen, T.E. Hansen, N. Lietaer, Nucl. Instrum. Meth. A **648**, 168 (2011)

[13] J.M. Verbeke, J. Randrup, R. Vogt, Comput. Phys. Commun. **222**, 263 (2018)

[14] A. Oberstedt, R. Billnert, F.J. Hambsch, S. Oberstedt, Phys. Rev. C **92**, 014618 (2015)

[15] S. Bjørnholm, J.E. Lynn, Rev. Mod. Phys. **52** (1980)

[16] V.V. Verbinski, H. Weber, R.E. Sund, Phys. Rev. C **7**, 1173 (1973)

[17] A. Chyzh, C.Y. Wu, E. Kwan, R.A. Henderson, T.A. Bredeweg, R.C. Haight, A.C. Hayes-Sterbenz, H.Y. Lee, J.M. O'Donnell, J.L. Ullmann, Phys. Rev. C **90**, 14602 (2014)

[18] F. Pleasonton, Nucl. Phys. A **213**, 413 (1973)

Dependence of the ROT effect on the energy of different light charged particles and on the incident neutron energy

Irina Guseva[1,*], *Alexei* Gagarski[1], *Friedrich* Gönnenwein[2], and *Yuri* Gusev[1]

[1]Petersburg Nuclear Physics Institute named by B.P. Konstantinov of NRC Kurchatov Institute, 188300 Gatchina, Russia

[2]Physikalisches Institut, Universität Tübingen, 72076 Tübingen, Germany

Abstract. The shift of the angular distribution of different light charged particles in ternary fission of ^{235}U induced by polarized neutrons, the so-called ROT effect, was estimated by modified trajectory calculations, which take into account the rotation of the compound nucleus. In previous publications only α-particles were considered. It is shown here that inclusion of tritons significantly improves the agreement of the energy dependence of the ROT effect with experiment while the inclusion of ^{5}He particles practically does not influence this dependence. In particular, the change in the magnitude of the ROT effect depending on the energy of incident neutrons is correctly predicted. Also, the ROT effect for gamma quanta and neutrons in binary fission is discussed along the same lines, because all mentioned effects are proportional to the effective angular velocity of the compound nucleus at the moment of scission.

1 Introduction

The ROT effect was discovered in 2005. It corresponds to a shift of the angular distribution of light charged particles (LCP), mainly alphas, which was observed in ternary fission of ^{235}U induced by cold polarized neutrons [1, 2]. It was assumed that this experimentally detected shift of the angular distribution is associated with the presence of rotational states in the level structure of the deformed compound nucleus. The validity of this hypothesis was confirmed by Monte-Carlo calculations [3, 4]. The rotational structure of the nucleus was known for a long time [5] but in case of unpolarized compound nuclei their rotations are arbitrarily oriented in space and can not to shift the angular distributions of ternary particles. In the present experiments the angular shift can arise only due to the spin polarization of the compound nucleus following capture of polarized neutrons. Flipping the spin of the polarized neutron beam inducing fission leads to opposite shifts Δ and $-\Delta$ of the LCP angular distribution, which allows to detect the ROT effect. The difference in angular distributions of LCP for the two spin orientations then is 2Δ.

*e-mail: guseva_is@pnpi.nrcki.ru

2 Experimental technique and data analysis

In Fig. 1 is shown the scheme of the experimental setup for the measurements to be discussed. The longitudinally polarized neutron beam was hitting the fissile target ^{235}U mounted at the center of a reaction chamber. Detectors for fission fragments (FFs) and LCP were installed in a plane perpendicular to the neutron beam.

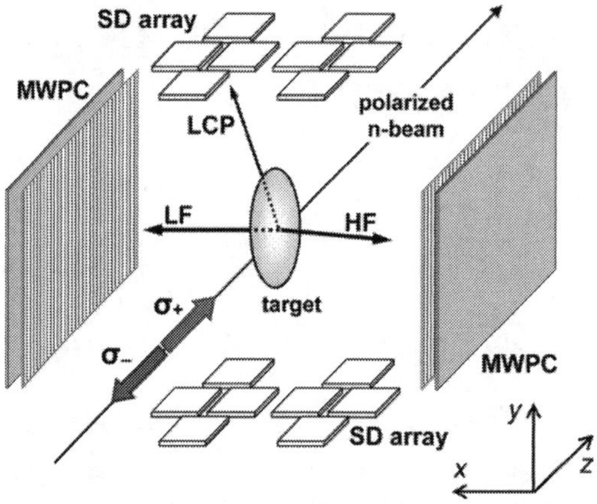

Figure 1. Scheme of the experimental setup: two multiwire proportional counters (MWPC) facing each other to the left and right of the target intercept fission fragments. LCPs are measured by two arrays of up to twelve Si detectors on the top and bottom of the target.

We assume that the ROT effect mostly develops after the nucleus rupture and when nuclear forces have already ceased to influence the trajectories of FFs and LCPs flying apart. This phenomenon then arises due to the motion of charged objects in the rotating field of Coulomb forces.

The time dependence of the motion for two fission fragments and a ternary particle due to their mutual Coulomb interaction can be reproduced using Monte-Carlo simulations [3, 4]. Initially it is necessary to perform calculations without considering the rotation of the fissile system [6]. Since this motion cannot be calculated in closed analytical form, it is necessary to replace the differential equations of motion

$$\frac{dX_{ij}}{dt} = V_{ij}, \quad m_i \frac{dV_{ij}}{dt} = F_{ij} \tag{1}$$

by a set of difference equations

$$X_{ij}^{n+1} = X_{ij}^{n} + \tilde{V}_{ij}^{n}\Delta t, \quad V_{ij}^{n+1} = \tilde{V}_{ij}^{n} + \frac{1}{2m_i}F_{ij}^{n}\Delta t, \tag{2}$$

where

$$\tilde{V}_{ij}^{n} = V_{ij}^{n} + \frac{1}{2m_i}F_{ij}^{n}\Delta t. \tag{3}$$

Here the subscript i denotes a particle with the masse m_i, while j corresponds to its component index for the coordinate X_{ij}, the velocity V_{ij} and the force F_{ij} acting on this particle. The superscript n refers to the value of the parameter after the n-*th* time interval. By iterations the time dependence of the coordinates for FFs and LCPs is found step by step.

Such standard trajectory calculations are commonly applied to describe spontaneous ternary fission or ternary fission induced by unpolarized neutrons [7, 8]. Our goal at this first stage of calculations is to find parameters of the nuclear system at a moment close to its rupture. It is necessary to determine mass distributions of FFs, positions of all partners with respect to the center of mass and initial velocity distributions of FFs and LCP. In the case of correctly fitted initial parameters of the considered system, the calculated final distributions for fission fragments and light charge particles should coincide with those that were experimentally recorded for a non-rotating fission system, in our case for ternary fission induced by unpolarized neutrons.

Found in this way, the initial configuration of the nuclear system allowed us to calculate its moment of inertia $\Im$ immediately after the nucleus rupture.

Then the initial angular velocity $\omega(J, K)$ can be obtained as the ratio of the mean value of the compound nucleus spin projection $< J_z(K) >$ onto the axis z of neutron beam polarization to the moment of inertia $\Im$:

$$\omega(J, K) =< J_z(K) > /\Im. \tag{4}$$

In the process of nuclear fission induced by slow polarized neutrons two spins of the compound nucleus may appear. They are equal to $J_+ = I + 1/2$ and $J_- = I - 1/2$, where I is the spin of the target nucleus.

The partial angular velocities $\omega_{+/-}(J, K)$ depend on the spins $J_{+/-}$, the projections $K_{+/-}$ on the fission axis, the moment of inertia $\Im$ and neutron beam polarization p_n [9]:

$$\omega_+(J, K) = +\frac{p_n\hbar}{2\Im} \cdot \frac{J_+(J_+ + 1) - K_+^2}{J_+}$$

$$\omega_-(J, K) = -\frac{p_n\hbar}{2\Im} \cdot \frac{J_-(J_- + 1) - K_-^2}{J_- + 1}. \tag{5}$$

The different signs of the angular velocities for J_+ and J_- indicate opposite directions of the fissile system rotation.

The ROT effect 2Δ in the angular distributions of ternary particles is proportional to the initial effective angular velocity ω_{eff}. This velocity is the sum of the two partial angular velocities ω_+ and ω_- for the two spins J_+ and J_- with coefficients depending on the partial fission cross sections:

$$\omega_{\text{eff}}(J, K) = \omega_+(J_+, K_+)\frac{\sigma_f(J_+)}{\sigma_f(J_+) + \sigma_f(J_-)} + \omega_-(J_-, K_-)\frac{\sigma_f(J_-)}{\sigma_f(J_+) + \sigma_f(J_-)}. \tag{6}$$

Based on the obtained effective angular velocity ω_{eff}, one can proceed to the second stage of trajectory calculations. In the present case the fissile system is rotating around the z-axis of neutron beam polarization in Fig. 1, and it is only necessary to modify the initial velocities of the objects described by taking into account their additional components tangential to this rotation. Considering spin-flip of the polarized neutron beam, this must be performed twice. The following iterative steps of the calculations are no different from the standard method. Using modified trajectory calculations one should recalculate all distributions.

As follows from the calculations (see Fig. 2), the rotational speed of the nuclear system after rupture is variable. It depends on the moment of inertia of the fissile system at any given time. This rotation starts with a huge angular velocity. It reaches about 10^{18} turns per second immediately after scission and tends to zero after about 5 zs.

As a result of the rapid decrease in the angular velocity, the fission fragments and the ternary particle can change their directions of motion only during a very short time, and then they move along straight lines. Already after a short time the angle of rotation of each reaction partner reaches its final value. In other words, each partner has got its final direction

https://doi.org/10.1051/epjconf/202125600006

of motion leading possibly to a detector for registration. In the experiment, the angle of α-particle is measured from the final direction of motion of the light fragment.

Figure 2 shows a typical example of time dependencies of deflection angles for LF and LCP. It is observed that already in a few zs the LCP trajectory lags behind the LF in turning angle. The final lag angle Δ is reached in only $\approx$5 zs. The angular LCP distribution, which is the result of many fission events, is hence shifted compared to a non-rotating composite system by a similar angle Δ, but which is averaged over all fission events.

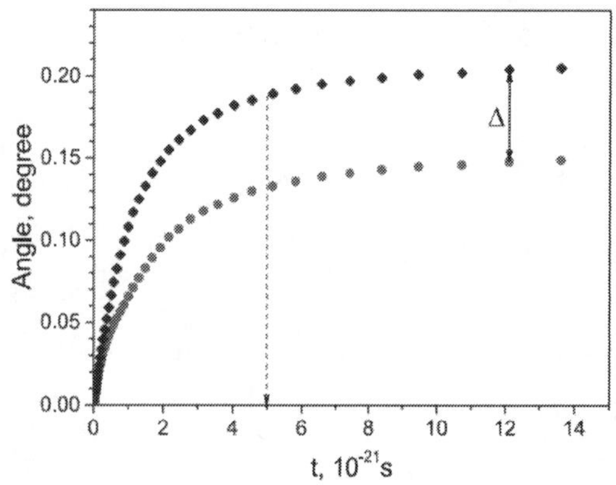

Figure 2. Time dependencies of deflection angles θ_α for α-particle (full dots) and θ_{LF} for LF (rhombs). $\Delta = \theta_{\mathrm{LF}} - \theta_\alpha$. Calculations were performed with $< J_z >= 1\,\hbar$.

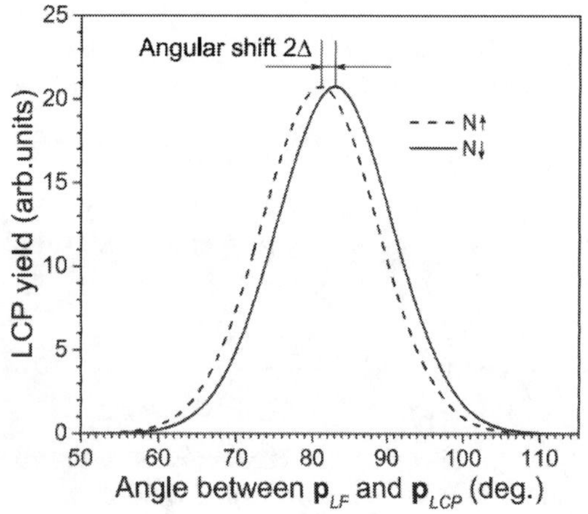

Figure 3. Double shift 2Δ of the angular distribution of ternary particles in (*n,f*) reactions depending on the orientation of neutron spin inducing fission. N↑ and N↓ are emission rates for neutron spin parallel and antiparallel to ($\mathbf{p}_{\mathrm{LF}} \times \mathbf{p}_{LCP}$), respectively. Here the angular shift is about 10 times greater than the experimental one for better visualization.

Taking into account the spin-flip technique, we need to repeat the last version of trajectory calculation, in which the directions of the additional velocities should be reversed. The difference in the angular distributions of ternary particles corresponding to different signs of the neutron beam polarization gives us the sought-for double shift 2Δ of this angular distribution (Fig. 3).

To get the best fit between experimental shift and the result obtained by modified trajectory calculations, which take into account rotation of the compound nucleus, we can change only the spin projections $K_{+/-}$ of the compound nucleus, while the other initial parameters of the fissile system were determined on the first standard stage of trajectory calculations. It was concluded that for the reaction ^{235}U(n, f) induced by cold neutrons combinations of the nuclear spin and its projection $(J_+, K_+) = (4, 0)$ and $(J_-, K_-) = (3, 2)$ are dominant.

3 Results of calculations

As already pointed out, originally these calculations have been performed only with alphas as the third particle, since they dominate in ternary fission. The evaluated angular shift averaged over the energy of α-particles was in good agreement with the experimental result, but the detailed distribution of the calculated ROT-effect values depending on the energy of ternary particles deviated from the experimental data. The experimental angular shift was larger than the calculated result in the energy range ($8 \div 13$) MeV of the third particle (Fig. 15 in [2]).

This discrepancy can be explained by the presence of other ternary particles besides alphas, since the corresponding measurements in this experiment were performed without precise identification of the ternary particle type. Accounting for the presence of tritons, which contribute about 7% to the LCP yields, it was possible to bring the calculated results in better agreement to experimental data. The results of such calculations were presented in [10].

However, besides α-particles and tritons, there are ternary particles of ^{5}He with a sizable contribution to ternary fission. Therefore, it was decided to start a new version of trajectory calculations including this isotope. We rely on ref [11], where the LCP yields in spontaneous fission of ^{252}Cf were studied. It was observed that in addition to "true" α-particles, which are formed during the nucleus separation, about 20% of the α-spectrum can be produced due to disintegration of ^{5}He into α-particle and a neutron. This is assumed to be true also in neutron induced fission of ^{235}U. Further, the fraction of emitted ^{5}He particles and their energy distribution for ternary fission of ^{235}U were taken to be identical to those for ^{252}Cf (see Table 1).

Evidently, it was necessary to take into account the very short lifetime of ^{5}He, which is approximately 1×10^{-21}s. This ternary particle rapidly decays into ^{4}He and a neutron. In addition, the smaller energy from ^{5}He decay should be considered when estimating the dependence of ROT effect on the energy of LCPs.

Table 1. Experimentally obtained energy distributions of LCP for spontaneous fission of ^{252}Cf [11].

LCP	<E>	TWHM
^{4}He	15.7(2)	10.9(2)
True ^{4}He	16.4(3)	10.3(3)
Residues from ^{5}He	12.4(3)	8.9(5)

Figure 4 shows the energy distributions for tritons, true α-particles and for isotope of ^{5}He, which were obtained by standard trajectory calculations. The initial parameters of the fissile system required for this task were chosen suitable to reproduce the energy distributions of ternary particles in the spontaneous fission of ^{252}Cf. The yields of emitted particles were taken in agreement with experimental data for neutron induced fission of ^{235}U and evaluations in ref. [11].

The modified trajectory calculations based on the adopted initial parameters and the calculated effective angular velocity allowed to find the angular shifts for tritons, true α-particles, and ^{5}He isotope decaying into an alpha and a neutron. The resulting ROT effect as a function of the kinetic energy of the LCPs is visualized in Fig. 5a.

By summing ROT angles 2Δ for the contributing LCPs with their respective weights, we obtain the result to be compared with the experimentally observed effect shown in Fig.5b. Evidently, the inclusion of tritons in the calculations gives a significant contribution to the energy dependence of the ROT effect at low LCP energies and convincingly improves agreement with experiment. On the other hand, the ROT effects for the ^{4}He and ^{5}He isotopes are virtually identical and the inclusion of the ^{5}He isotope cannot bring a noticeable change compared with the previous description of its energy dependence [10].

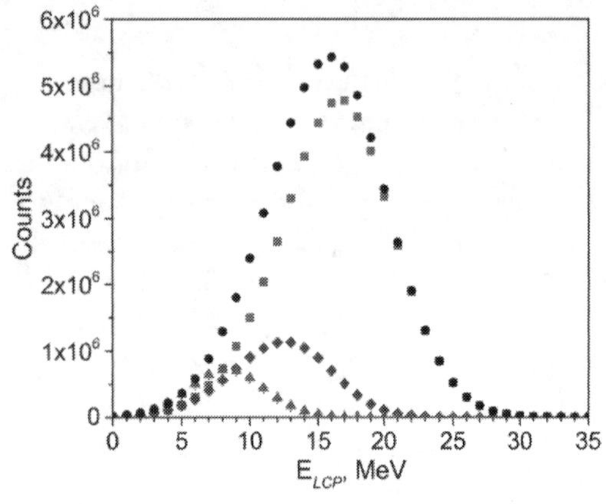

Figure 4. The calculated LCPs energy distributions: for ^{3}H (triangles), for true α-particles (squares), for ^{5}He, which turns into α-particle plus a neutron (rhombs), and for the sum of true and residual α-particles (full dots).

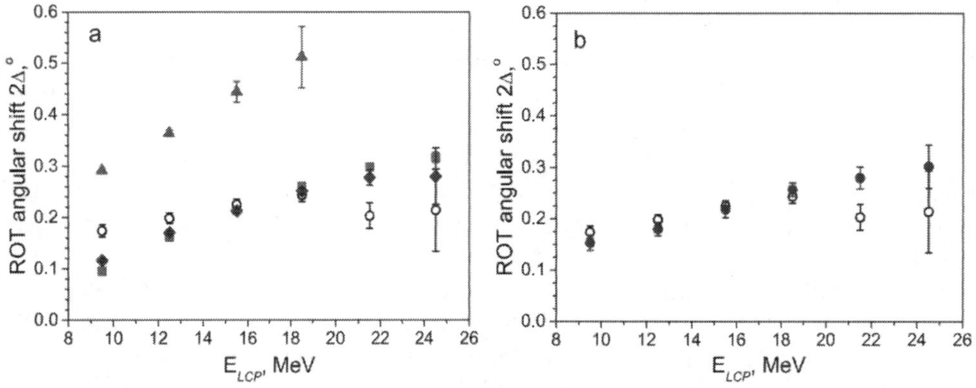

Figure 5. a) Calculated ROT effect for ^{3}H (triangles), true ^{4}He (squares), ^{5}He which turns into an α-particle plus a neutron (rhombs) and experimental data (opened dots); b) the calculated ROT for the sum of all LCP components considered (full dots) in comparison with experimental data (open dots).

4 Dependence of the ROT effect on the incident neutron energy

As discussed in connection with eq. (5), the rotational speeds for the two possible spins of the ^{236}U* compound nucleus, $J_+ = 4$ and $J_- = 3$ have different signs. Both contribute to the effective angular velocity ω_{eff} in Eq. (6) as the start velocity for trajectory calculations and ensuing ROT effects. But the sign and size of the effective angular velocity and, therefore, the sign and magnitude of the ROT effect also depends on the partial fission cross-sections, more precisely, on their ratio. This ratio varies with the energy of the incident neutron.

The energy dependence of the spin-separated fission cross sections for neutron induced fission of ^{235}U were calculated by the computer code SAMMY (see Fig. 6). This figure shows two curves for the partial fission cross sections and their sum for the target of ^{235}U. In these calculations the data of resonance parameters were taken from ENDF/B-VIII.0 (USA, 2018) [12].

Experiments in ternary fission were performed in the energy region of cold neutrons. Here, the partial fission cross section for $J = 4$ is approximately 2 times larger than the corresponding value for $J = 3$. Furthermore, as follows from Eq. (5), the positive partial angular velocity of the state $(J_+, K_+) = (4, 0)$ exceeds in absolute value the negative rotational speed of the compound state $(J_-, K_-) = (3, 2)$. As a result, we get in this region a positive and sizable value of the ROT effect.

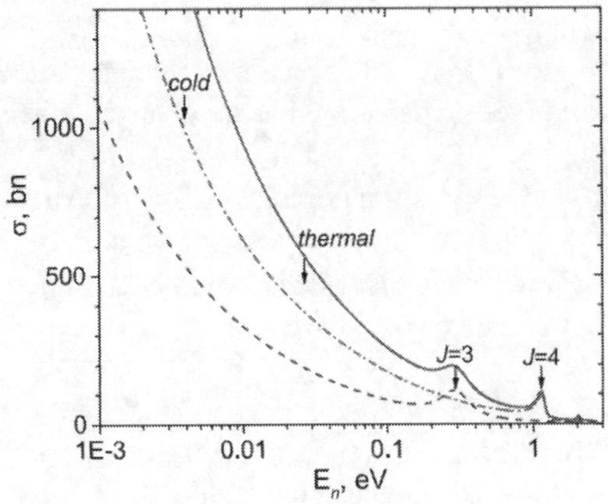

Figure 6. Spin-separated fission cross-sections for the target of ^{235}U as a function of the energy of the incident neutron: dashed line for $J = 3$, dash-doted line for $J = 4$, solid line for the sum of both.

Practically the same results should be expected for thermal neutrons. Closer to the resonance at $E_n = 0.3$ eV the situation is different. The spin of this resonance is defined as $J = 3$. Although at its top the partial fission cross section is slightly larger than the value corresponding to spin $J = 4$, the rotation in the positive direction continues to dominate due to the significant larger absolute value of the positive partial angular velocity compared to the negative velocity. As a result, the sign of the ROT effect remains also here positive, but its size becomes two times less than for thermal or cold neutrons. For the resonance region near 1 eV, on the contrary, we can expect an increase in the ROT effect compared to the cold or thermal region by a factor 2.3.

5 Summary and outlook

In the present semi-classical interpretation of the ROT effect, the main cause of the shift in the angular distribution of light charged particles observed in the ternary fission induced by polarized neutrons is the interaction of charged objects when they move in a rotating Coulomb field. It is especially important to take into account the development of this phenomenon in time, namely, that the angular velocity very quickly tends to zero. If case the angular velocity would remain constant for a long period of time, no experiments of this type could have predictable results, since this would lead to a dependence of the magnitude of the effect on the distance between the detectors and the target.

The evolution in time of the motion of three charged objects in a Coulomb field can well be described classically. The quantum-mechanical aspect of the ROT effect phenomenon is represented by the initial angular velocity ω_{eff}. Moreover, using the values of partial fission cross sections calculated by the computer code SAMMY based on the most recent Nuclear Data Files, the interference of s-resonances with the same spins is duly described.

The need to take into account the interference of s-resonances with different spins, on which the authors [13] insist to describe the ROT effect, cannot be considered proven. Justifying the correctness of their approach, they do not perform real calculations in accordance with their formulas, but simply fitted the corresponding coefficients to the experimental ones. In these calculations, the resonance parameters are not used at all, not to mention the interference between them.

In contrast to the authors of [13], we obtained real parameters of s-resonances from the Evaluated Nuclear Data File in a fairly wide energy range and used them to correctly calculate spin-separated fission cross sections. We need them in order to determine the ratio of rotations of compound-nuclei in two opposite directions. In addition, we can quite adequately estimate the moment of inertia of the fissile system from the point of its rupture to the end of rotation, which the authors [13] do not have. They do not at all follow the change in time of the size of the rotating system, which is characterized by the distances between interacting charged objects and determines the moment of inertia of interest to us, as well as the change in their kinematic parameters.

Our evaluation of the magnitude of the ROT effect occurs independently of the experimental results and only after performance of calculations are these values compared.

As was shown, the result of our calculations for the magnitude and sign of the ROT effect are in good agreement with experimental data, both averaged over the energy of LCPs as well as their detailed LCP energy dependence.

At present, all measurements of the ROT effect in ternary fission are carried out only at incident neutrons from the cold energy region. To test our predictions about the dependence of the ROT effect on the neutron energy, it should be of great interest to measure this effect at neighboring *s*-resonances.

It should finally be noted that this predicted energy dependence is also valid for the ROT effect for neutrons and gammas in binary fission [14], because these effects are likewise proportional to the effective angular velocity of the compound nucleus at the moment of scission. These conclusions can be tested in experiment by comparing the ROT effect values in the resonance energy region [15] with those obtained for cold or thermal neutrons.

References

[1] F. Gönnenwein, M. Mutterer, A. Gagarski, I. Guseva, G. Petrov, V. Sokolov, T. Zavarukhina, Yu. Gusev, J. von Kalben, V. Nesvizhevski, and T. Soldner, Phys. Lett. B **652**, 13 (2007)

[2] A. Gagarski, F. Gönnenwein, I. Guseva, P. Jesinger, Yu. Kopatch, T. Kuzmina, E. Lelièvre-Berna, M. Mutterer, V. Nesvizhevsky, G. Petrov, et al., Phys. Rev. C **93**, 054619 (2016)

[3] I.S. Guseva and Yu.I. Gusev, Bull. Russ. Acad. Sci. Phys. **71**, 367 (2007)

[4] I. Guseva and Yu. Gusev, AIP Conf. Proc. **1175**, 355 (2009)

[5] A. Bohr and B.R. Mottelson, Nuclear Structure, vol. 2: Nuclear Deformations (Benjamin, New York, 1975)

[6] Y. Bohne, Z. Fraenkel and I. Nebenzahl, Phys. Rev. **156**, 1305 (1967)

[7] M. Borkovski, Yu. Gusev, Yu. Zalite, and D. Seliverstov, Proceedings of the 18th International Symp. on Nucl. Physics, Physics and Chemistry of Fission, Gaussing (1988), p. 181

[8] W. Baum, Ph.D. Thesis, Institute für Kern-physik, TH Darmstadt (1992)

[9] V.E. Bunakov and S.G. Kadmensky, Phys. At. Nucl. **71**, 1200 (2008)

[10] I. Guseva, A. Gagarski, F. Gönnenwein, and Yu. Gusev, in Neutron Spectroscopy, Nuclear Structure, Related Topics: Proceedings of the XXV International Seminar on Interaction of Neutrons with Nuclei, Dubna, May 22-26, 2017 (Joint Institute for Nuclear Research, Dubna, Russia, 2018), p. 355

[11] Yu.N. Kopatch, M. Mutterer, D. Schwalm, P. Thirolf, and F. Gönnenwein, Phys. Rev. C **65**, 044611 (2002)

[12] D.A. Brown, M.B. Chadwick, R. Capote, A.C. Kahler, A. Trkov, M.W. Herman, A.A. Sonzogni, Y. Danon, A.D. Carlson, M. Dunn et al., Nucl. Data Sheets **148**, 1-142 (2018)

[13] S.G. Kadmensky, D.E. Lyubashevsky, and P.V. Kostryukov, Phys. At. Nucl. **82**, 267 (2019)

[14] I. Guseva, A. Gagarski, Yu. Gusev, G. Petrov, G. Valski, Phys. Part. Nucl. Lett. **10**, 351 (2013)

[15] Yu. Kopatch, V. Novitsky, G. Ahmadov, A. Gagarsky, D. Berikov, G. Danilyan, V. Hutanu, J. Klenke, and S. Masalovich, EPJ Web Conf. **169**, 00010 (2018)

The description of the excitation energy sharing in nuclear fission within the Langevin approach

F.A. Ivanyuk[1,2,*] and *S.* Chiba[2,**]

[1]Institute for Nuclear Research, 03028 Kiev, Ukraine
[2]Tokyo Institute of Technology, 152-8550 Tokyo, Japan

Abstract. We apply the four-dimensional Langevin approach to the description of fission of ^{235}U by neutrons and calculate the dependence of the excitation energy of fission fragments on their mass number. For this we run the Langevin equations until the compound nucleus splits into two separated fragments. This is possible since the we used in this work two-center shell model shape parametrization that describes well both compact and separated shapes. The excitation energies of each fragment are calculated assuming that the temperatures of both fragments are the same. The deformation energy of the fragment immediately after scission is added to its excitation energy. The saw-tooth structure of the dependence neutron multiplicity on the fragment's mass number in reaction ^{235}U + n at E_n = 5 Mev is qualitatively reproduced.

1 Introduction

One of the most successful approaches to the description of fusion-fission reactions and the fission process is the approach based on the Langevin equations for the shape degrees of freedom. In this approach instead of co-ordinates of individual nucleons one introduces the parametrization of the shape of nuclear surface and solves the dynamical equations for the time evolution of deformation parameters and conjugated momenta. The internal state of the system in such calculations is assumed not to deviate much from the thermal equilibrium.

With the code developed at Tokyo Institute of Technology [1–5] during last few years we have managed to reproduce rather accurately the mass and total kinetic energy distribution of fission fragments. We have also found a very clear correlation between the neutron multiplicity and the deformation of fission fragment just before scission.

The existing descriptions of the sharing of excitation energy [4, 6–12] are based on very different assumptions and approximations that even contradict to each other. So, the reason of the saw-tooth structure at present is not clear. In this work we exploit the found correlation between the neutron multiplicity and the deformation of fission fragment in order to clarify the reason for the saw-tooth structure of neutron multiplicity and its dependence on the excitation energy.

*e-mail: ivanyuk@kinr.kiev.ua
**e-mail: chiba.satoshi@nr.titech.ac.jp

2 The model

We describe the fission process within the Langevin approach [13], i.e., by solving the equations for the time evolution of the shape of nuclear surface of fssioning system. For the shape parametrization we use that os two-center shell model (TCSM) [14] with 4 deformation parameters $q_\mu = (z_0/R_0, \delta_1, \delta_2, \alpha)$. Here z_0/R_0 refers to the distance between the centers of left and right oscillator potentials with $R_0 = 1.2\sqrt[3]{A}$ being the radius of spherical nucleus with mass number A. The parameters $\delta_i = 3(a_i - b_i)/(2a_i + b_i)$, where i={1,2}, describe the deformation of the right and left fragment tips. The fourth parameter q_4 is the mass asymmetry $\alpha = (V_1 - V_2)/(V_1 + V_2)$ where V_1 and V_2 are the volumes to the left and right from the point $z = 0$. The fifth parameter of TCSM shape parametrization ϵ, defined as the ratio of actual and oscillator potentials at $z = 0, \epsilon \equiv E/E_0$, is kept constant, $\epsilon = 0.35$, in all our calculations.

The first order differential equations (Langevin equations) for the time dependence of collective variables q_μ and the conjugated momenta $p_{\rm mu}$ are:

$$\frac{dq_\mu}{dt} = \left(m^{-1}\right)_{\mu\nu} p_\nu, \qquad (1)$$
$$\frac{dp_\mu}{dt} = -\frac{\partial F(q,T)}{\partial q_\mu} - \frac{1}{2}\frac{\partial m^{-1}_{\nu\sigma}}{\partial q_\mu} p_\nu p_\sigma - \gamma_{\mu\nu} m^{-1}_{\nu\sigma} p_\sigma + g_{\mu\nu} R_\nu(t),$$

where the sums over the repeated indices are assumed. In Eqs. (1) $F(q,T)$ is the temperature dependent free energy of the system, and $\gamma_{\mu\nu}$ and $(m)^{-1}_{\mu\nu}$ are the friction and inverse of mass tensors and $g_{\mu\nu}R_\nu$ is the random force.

The free energy $F(q,T)$ is calculated as the sum of liquid drop deformation energy and the shell correction $\delta F(q,T)$. The damping of shell correction $\delta F(q,T)$ with the excitation energy is described in detail in [15]. The single particle energies are calculated with the deformed Woods-Saxon potential [16–18] fitted to the aforementioned TCSM shape parameterizations. The collective inertia tensor $m_{\mu\nu}$ is calculated within the Werner-Wheeler approximation [19] and for the friction tensor $\gamma_{\mu\nu}$ we used in present work the wall-and-window formula [20, 21].

The random force $g_{ij}R_j(t)$ is the product of white noise $R_j(t)$ and the temperature dependent strength factors g_{ij}. The factors g_{ij} are related to the temperature and friction tensor via the modified Einstein relation [22],

$$g_{ik}g_{kj} = T^*\gamma_{ij}\,, \qquad \text{with } T^* = \frac{\hbar\varpi}{2}\coth\frac{\hbar\varpi}{2T}\,, \qquad (2)$$

where T^* is the effective temperature. The parameter ϖ is the local frequency of collective motion [22]. The minimum of T^* is given by $\hbar\varpi/2$, which corresponds to the zero point energy of oscillators forming the heat bath. Based on the pioneering works [23, 24], we estimated the zero point energy as 1 MeV, which lies in the middle of the corresponding quantities for various modes 0.45 to 2.23 MeV estimated in Ref. [24]. The temperature T in this context is related to the initial excitation energy E_x and the internal energy $E_{\rm int}$ by,

$$E_{\rm int} = E^* - \frac{1}{2}\left(m^{-1}\right)_{ij} p_i p_j - V(q, T=0) = aT^2\,, \qquad (3)$$

where a is the level density parameter.

More details are specified in our earlier publications [1–3].

Initially, the momenta p_i were set to zero, and calculations are started from the ground state deformation. Such calculations are continued until the trajectories reach the "scission point", which was defined as the point in deformation space with a certain value of neck radius.

3 Numerical results

A single event of Langevin calculation typically begins from the ground state deformation. If it fails to achieve scission configuration after 10,000 fm/c, the calculation is terminated. Usually 500,000 events are done per nuclei but we occasionally increase the number of events if it was a critical calculation. Typically, the trajectories stay for a lot of time inside that

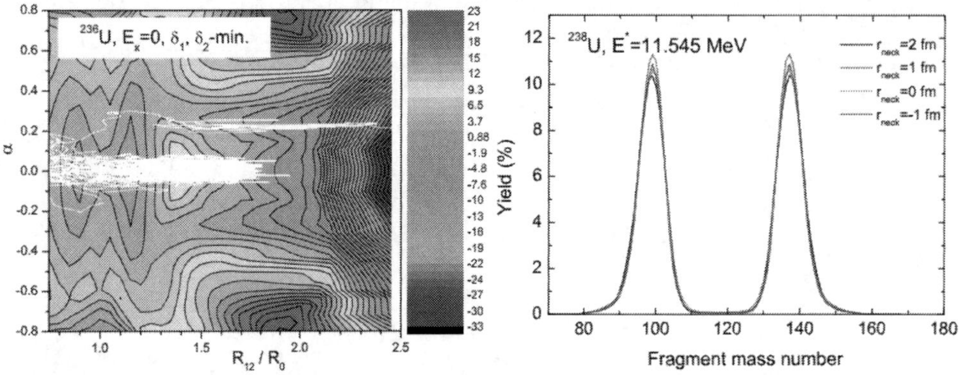

Figure 1. *Left*: the potential energy of ^{236} in elongation R_{12} and mass asymmetry α coordinates, minimized with respect to δ_1 and δ_2 deformation parameters. White line is an example of trajectory in deformation space. *Right*: the mass distribution of fission fragments calculated at different critical value of the neck parameter, $r_{\rm neck}$ = 2, 1 0 and −1 fm.

saddle, some of them go over the saddle and reach the scission point, see Fig. 1a. In our previous calculations we defined the scission point as a deformation at which the neck radius turns into zero. However there are some indications that the fissioning nucleus breaks apart at rather thick neck. This was shown a long ago within the liquid drop model [25]. The calculations within Time-Dependent Hartree-Fock approach are also stopped at the finite value of the neck of the order of 1–2 fm.

In Fig. 1b we compare the mass distribution of fission fragment of ^{236}U at the excitation energy E^* = 11.545 MeV. As one can see, the mass distributions are rather stable with respect to variation of $r_{\rm neck}$ within a reasonable limits, 2 fm ≤ $r_{\rm neck}$ ≤ −1 fm. The negative neck radius refer here to the separated fragments, see Fig. 2.

In our previous publication [4] we stop the calculations when the neck radius turns into zero. Then we fitted this just before scission shape by the shape of two separated fragments and calculated the deformation and excitation energies of each fragments separately. Since the scission point is not well defined, in present work we avoid the fitting of just before scission shape and continue Langevin calculations until the compact shape would split into two fragments. This is possible since the TCSM shape parametrization describes well both compact and separated shapes. Practically, we run calculations until the neck parameter became negative, $r_{\rm neck}$ = − 1 fm. Then for each fragment we calculated the deformation and excitation energies. The deformation energy was calculated by the macroscopic-microscopic method, $E_{\rm def}(q) = E_{\rm LDM}(q) + \delta E_{\rm shell}(q) + \delta E_{\rm pair}(q)$, the same as for the compact configuration, and for the thermal excitation energy we used the BCS-approximation,

$$E^*(T) = E_{\rm BCS}(q^{\rm (ias)}, T) - E_{\rm BCS}(q^{\rm (ias)}, T = 0) \quad (4)$$

Here $q^{\rm (ias)}$ is the set of deformation parameters "immediately after scission" (at $r_{\rm neck}$ = −1 fm). The temperature of light and heavy fragments immediately after scission was assumed to be the same and was found from the requirement that the sum of excitation energies

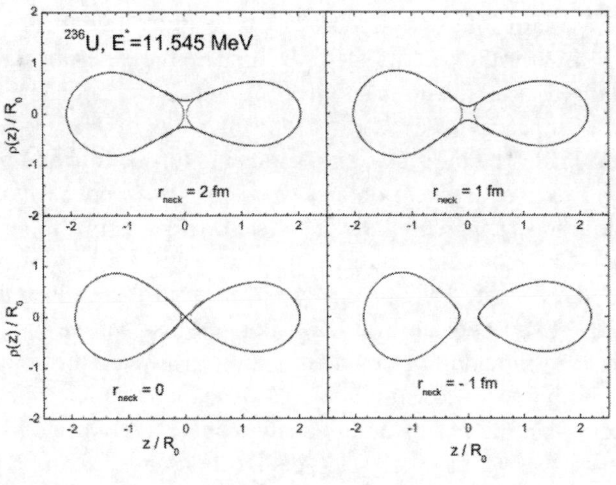

Figure 2. The shape of fissioning ^{236}U at critical neck radius $r_{\text{neck}} = 2, 1, 0$ and -1 fm.

of light and heavy fragments is equal to the total excitation energy E_{int}, see Eq. (3), resulting from the Langevin calculations,

$$E_L^*(T) + E_R^*(T) = E_{\text{int}}. \tag{5}$$

In Fig. 3a we plot separately the ground state energy E_{gs}, the deformation energy $E_{\text{def}}^{(\text{ias})}$ and the excitation energy E^* as function of fragment mass number. One can see that the main

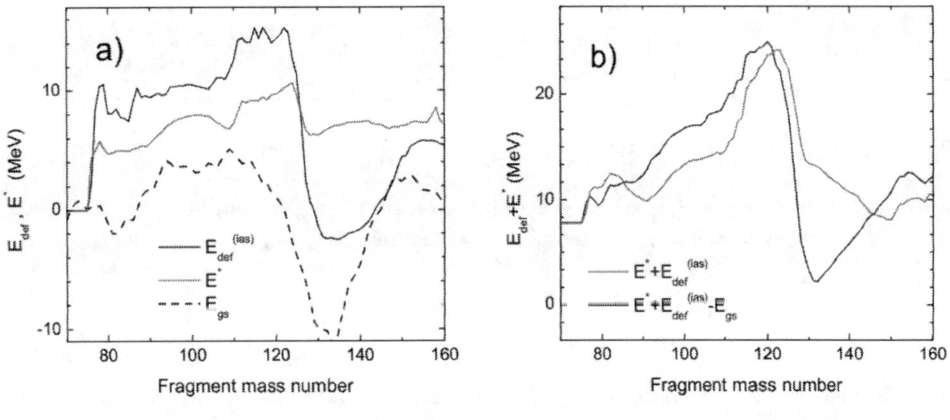

Figure 3. a) The average value of the ground state energy E_{gs} (dash), deformation $E_{\text{def}}^{(\text{ias})}$ (black) and excitation E^* (red) energies of fission fragments after the neck rupture as function of the fragment mass number; b) the sum $E^* + E_{\text{def}}^{(\text{ias})}$ (black) and the total excitation energy $E^* + E_{\text{def}}^{(\text{ias})} - E_{\text{gs}}$ as function of the fragment mass number.

contribution of the saw-tooth structure of excitation energy comes mainly from the deformation energy of fission fragment. This result is not new. It was noticed by Wilkins et al [26] already in 1976 that the dependence of quadrupole moment os fission fragment on its mass number is very similar to that of neutron multiplicity. The sharing of excitation energy seems not so important. Even if one would share the excitation equally between the fragments, or share it proportionally to the mass of fragment, the saw-tooth structure would remain due to $E_{\text{def}}^{(\text{ias})}$.

The sum $E^* + E_{\text{def}}^{(\text{ias})}$ surprisingly well reproduces the structure of neutron multiplicity, see the left part of Fig. 4. In the right part of Fig. 4 we compare the experimental values of average neutron multiplicity $\bar{\nu}$ from [27], with that derived from the excitation energy $E^* + E_{\text{def}}^{(\text{ias})}$. For $\bar{\nu}$ we have calculated the decay of excited fission fragments by Hauser-Feshbach theory as implemented in Ref. [28]. The transmission coefficients for neutron and proton emission (if any) were calculated by the spherical optical model by using optical potential parameters suggested by Koning and Delaroche [29]. The level densities of the residual nuclei were calculated by Gilbert-Cameron composite formula with parameters given in Ref. [30]. At low excitation energies, the discrete level data were taken from the evaluated nuclear structure data stored in RIPL-3 [31] (updated in 2012). For γ emission, we have included E1, E2 transitions, and M1 spin-flip and scissors modes. We have fixed the spin and parity of the decaying fragment in this work for simplicity. We tried calculation with $J^\pi = 0^+$ for even and $\frac{1}{2}^+$ for odd nuclei, but the results do not change appreciably if we increase spin values by 8 $\hbar$. We recognize that we have to assign distributions of spin and parity as performed in Ref. [28] for final calculation , but a "fixed" spin-parity treatment should be enough as an initial investigation.

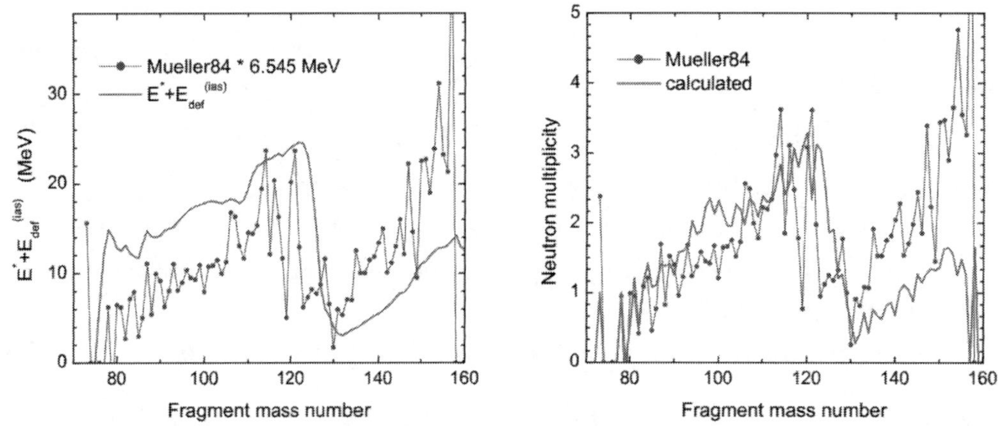

Figure 4. *Left*: the sum of the excitation and the deformation energy (red) in comparison with the experimental value of $\bar{\nu}$ multiplied by the neutron separation energy. *Right*: the comparison of the calculated and experimental values of $\bar{\nu}$.

As one can see from the left part of Fig. 4, the saw-tooth structure of $\bar{\nu}$ is well reproduced. The difference between the measured and calculated $\bar{\nu}$ for light fission fragments is within the experimental uncertainties. For the heavy fragments the calculated $\bar{\nu}$ is somewhat smaller.

Here we should note that $E^* + E_{\text{def}}^{(\text{ias})}$ is only part of excitation energy. It is believed that the prompt neutrons are emitted from completely accelerated fragments, thus, from the fragments at their ground state deformation. The total excitation energy available for the neutron emission is then $E^* + E_{\text{def}}^{(\text{ias})}$ minus the ground state energy E_{gs}, $E_{\text{tot}}^* = E^* + E_{\text{def}}^{(\text{ias})} - E_{\text{gs}}$. The subtraction of E_{gs} destroys significantly the similarity of $\bar{\nu}$ and E_{tot}^*. The minimum of E^* moves from $A_F \approx 132$ to $A_F \approx 150$. Consequently, the agreement between experimental and calculated values of $\bar{\nu}$ became worth, see Fig. 5.

We do not see what might be wrong in our approach. Once we got the shape of separated fragments from output of Langevin approach, the calculating of deformation and excitation energies is done exactly, without any approximation, no fitted parameters like levels density parameter or damping of the shell correction. All the important properties of fissioning system, the shell effects, the pairing effects, the dependence of both on the excitation energy, are

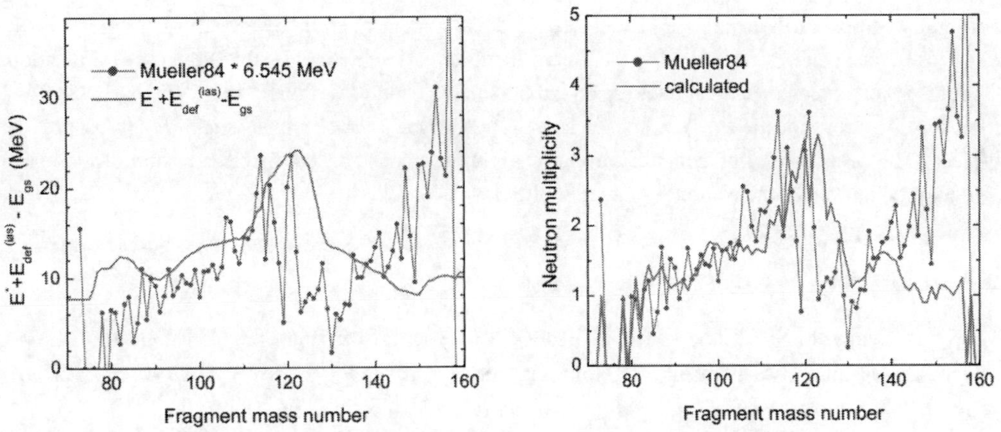

Figure 5. *Left*: the sum of the excitation E^* and the extra deformation energy $E_{\rm def} - E_{\rm gs}$(red) in comparison with the experimental value of $\bar{\nu}$ multiplied by the neutron separation energy. *Right*: the comparison of the calculated and experimental values of $\bar{\nu}$.

taken into account. In our opinion the similar problem with the subtraction of ground state energy should have other authors who explain the saw-tooth structure of $\bar{\nu}$.

In this work we calculate $E_{\rm def}^{\rm (ias)}$ for each fragment with respect to the macroscopic energy at the spherical shape. After the neck rupture the fragments relax to the spherical shapes (because they are still excited) within the very short relaxation time, $\tau \approx 5 \times 10^{-22}$ sec. Hence, what we calculate is the neutron emission in the very early stage of acceleration. Since we explain well the data one can infer that the neutron emission takes place shortly after scission.

The fact that the prompt neutrons are registered at the same direction as the fission fragment does not say much at what moment the neutron are actually emitted. Some part of neutrons may be emitted at the acceleration stage, before they relaxed to their ground state shape. One should keep in mind also the possibility of scission neutrons [32] which are emitted during the neck rupture, thus, from highly deformed fragments. The number of scission neutrons is not so small, it can reach one half of all prompt neutrons [33].

4 Summary

We used the output of the 4D Langevin approach to calculate the excitation energies of fission fragments in reaction ^{235}U + n at neutron energies E_n = 5 MeV. The Langevin code was run until the compact system would split into tao fragments.

The temperature of fission fragments was assumed to be the same and was found from the requirement that the sum of excitation energy of fission fragments immediately after scission is equal to the excitation energy of fissioning nucleus. We have found that the dependence of excitation plus deformation energy on the fragment mass number surprisingly well reproduces the saw-tooth structure of average neutron multiplicity $\bar{\nu}$. This can be considered as a hint that a considerably part of prompt neutrons is emitted not from the fully accelerated fragments but on the acceleration stage or during the neck rupture. The subtraction of ground state deformation energy from fragments excitation energy makes the agreement with experimental data much worth.

This study comprises the results of "Research and development of an innovative transmutation system of LLFP by fast reactors" entrusted to the Tokyo Institute of Technology by the Ministry of Education, Culture, Sports, Science and Technology of Japan (MEXT) and KAKENHI Grant Number 18K03642 from Japan Society for the Promotion of Science (JSPS). One of us (F.I.) was supported in part by the project "The investigation of statistical properties of collective states and nuclear fission of rear-earth and actinide nuclei" of the National Academy of Sciences of Ukraine.

References

[1] F.A. Ivanyuk, S. Chiba, and Y. Aritomo, Jour. Nucl. Sci. Tech. **53**, 737 (2016)
[2] M.D. Usang, F.A. Ivanyuk, C. Ishisuka, and S. Chiba, Phys. Rev. C **94**, 044602 (2016)
[3] C. Ishizuka, M.D. Usang, F.A. Ivanyuk *et al.*, Phys. Rev. C **96**, 064616 (2017)
[4] F.A. Ivanyuk, C. Ishizuka, M.D. Usang, and S. Chiba, EPJ Web of Conferences **169**, 00005 (2018)
[5] M.D. Usang, F.A. Ivanyuk, C. Ishisuka, and S. Chiba, Scientific Reports **9**, 1525 (2019)
[6] K.-H. Schmidt and B. Jurado, Phys. Rev. Lett. **104**, 212501 (2010)
[7] K.-H. Schmidt and B. Jurado, Phys. Rev. C **83**, 061601(R) (2011)
[8] C. Morariu, A. Tudora, F.-J. Hambsch, S. Oberstedt, C. Manailescu, J. Phys. G. Nucl. Part. Phys. **39**, 055103 (2012)
[9] B. Becker, P. Talou, T. Kawano, Y. Danon, and I. Stetcu, Phys. Rev. C **87**, 014617 (2013)
[10] A. Tudora, F.-J. Hambsch, I. Visana, G. Giubega, Nucl. Phys. A **940**, 242 (2015)
[11] M. Albertsson, B.G. Carlsson, T. Dossing, P. Moller, J. Randrup and S. Aberg, preprint nucl-th/1811.02283 (2018)
[12] A. Bulgac, S. Jin, K.J. Roche *et al.*, Phys. Rev. C **100**, 034615 (2019)
[13] Y. Abe, S. Ayik, P.-G. Reinhard and E. Suraud, Phys. Rep. **275**, 49 (1996)
[14] J. Maruhn and W. Greiner, Z. Phys. **251**, 431 (1972)
[15] F.A. Ivanyuk, C. Ishizuka, M.D. Usang and S. Chiba, Phys. Rev. C **97**, 054331 (2018)
[16] V.V. Pashkevich, Nucl. Phys. A **169**, 275 (1971)
[17] V.V. Pashkevich, Nucl. Phys. A **477**, 1 (1988)
[18] V.V. Pashkevich, A.Ya. Rusanov, Nucl. Phys. A **810**, 77 (2008)
[19] K.T.R. Davies, A.J. Sierk, J.R. Nix, Phys. Rev. C **13**, 2385 (1976)
[20] J. Blocki, Y. Boneh, J.R. Nix *et al*, Ann. Phys. **113**, 330 (1978)
[21] A.J. Sierk and J.R. Nix, Phys. Rev. C **21**, 982 (1980)
[22] H. Hofmann, D. Kiderlen, Int. Jour. Mod. Phys. E **7**, 243 (1998)
[23] N. Bohr and A.J. Wheeler, Phys. Rev. **56**, 426 (1939)
[24] D.L. Hill and A.J. Wheeler, Phys. Rev. **89**, 1102 (1952)
[25] V.M. Strutinsky, N. Ya. Lyashchenko, N.A. Popov, Nucl. Phys. **46**, 659 (1963)
[26] B.D. Wilkins, E.P. Steinberg, R.R. Chasman, Phys. Rev. C **14**, 1832 (1976)
[27] R. Müller, A.A. Naqvi, F. Käppeler, and F. Dickmann, Phys. Rev. C **29**, 885 (1984)
[28] S. Okumura *et al.*, J. Nucl. Sci. Tech., **55**, 1009 (2018)
[29] A.J. Koning and J.P. Delaroche, Nucl. Phys. **A 713**, 231 (2003)
[30] T. Kawano, S. Chiba and H. Koura, Nucl. Sci. Tech., **43**, 1 (2016)
[31] R. Capote *et al.*, Nucl. Data Sheets **110**, 3107 (2009)
[32] N. Carjan and M. Rizea, Phys. Rev. C **99**, 034613 (2019)
[33] N. Carjan, Contribution to Theory-5 Workshop

The ν-Ball campaign at ALTO: Study for neural network based trigger for fission process

Matthieu Lebois[1,2,*], *Nikola* Jovančević[3], *Damien* Thisse[1], *Jonathan* Wilson[1], *Rhiann* Canavan[4,5], *Mathias* Rudigier[4], and the N-SI-109 experimental team

[1]Université Paris-Saclay, CNRS/IN2P3, IJCLab, 91405 Orsay, France
[2]Institut Universitaire de France, 1 Rue Descartes, 75005 Paris, France
[3]Department of Physics, Faculty of Science, University of Novi Sad, Trg Dositeja Obradovica 3, 21000 Novi Sad, Serbia
[4]Department of Physics, University of Surrey, Guildford, GU2 7XH, UK
[5]National Physical Laboratory, Teddington, UK

Abstract. A γ-spectroscopy campaign named "ν-Ball" was perfomed at the ALTO facility. A large fraction of the beam time was dedicated to the fast neutron induced fission of two fissioning systems: ^{232}Th and ^{238}U. During the data analysis, it was noticed that the high activity of the ^{nat}Th was heavily contaminating any coincidence matrices (or cubes) built. This caused the identification of weakly produced fission fragments identification to be almost impossible. It was decided to explore the opportunity opened by new analysis methods based on neural networks algorithms. In this paper, the methods to build an adequate neural network and the results obtained for fission event reconstruction are presented.

1 Introduction

In 2017-2018, the ALTO facility of the "Laboratoire de Physique des 2 infinis Irène Joliot-Curie" (former "Institut de Physique Nucléaire d'Orsay") hosted an experimental campaign using the γ spectrometer named: ν-ball. Generally, γ-ray spectroscopy is a tool used in nuclear physics research, which has permitted the gathering of great quantities of information about excited nuclear states. Over the past decade, highly detailed coincidence spectroscopy has been performed over a wide range of atomic nuclei, allowing detection of γ transitions as weak as 10^{-5}[1–4] of the total reaction cross section. Information on spins and parities of these excited states can also be deduced with pure germanium detector arrays using measurement techniques such as γ-ray angular distributions, angular correlations and polarisations. Nowadays, development of fast-scintillator such as $LaBr_3$ allows the access to lifetime information in the 10 ps to 10 ns range, which is complementary to germanium based lifetime measurements techniques (over a 20 ns range) or plungers based measurement – to get lifetimes in the 1–100 ps range [5, 6]. In a similar way as the ROSPHERE array [7], the ν-Ballspectrometer has been implemented at the ALTO facility.

*e-mail: lebois@ijclab.in2p3.fr

Figure 1. Photograph of the ν-Ballarray installed in the experimental area of the ALTO facility (left panel) and a technical drawing of one ν-Ballhemisphere (right panel).

2 Presentation of the ν-Ballspectrometer & the ν-Ballcampaign

The ν-Ballarray combines 24 Ge clovers from the GAMMAPOOL [8], 10 co-axial high purity germanium (HPGe) detectors and 20 $LaBr_3$ or 34 PARIS phoswich detectors [9–11]. Figure 1 shows a photo of the array (on the left). On the right a technical drawing shows the three types of detector: HPGe clover detectors and their associated Compton shield (the two central rings), co-axial HPGe detectors (right ring) and $LaBr_3$ in the left ring.

To optimize ν-Ballperformances, a special attention has been brought to:

- Compton rejection *via* the use of Compton shields. In addition, heavy metal collimators on the BGO detectors were removed to obtain an energy measurement for the event-by-event calorimetry of the nuclear reactions studied.
- Inclusion of $LaBr_3$ detectors to facilitate measurements of excited nuclear states lifetimes in the 10 ps – 10 ns range.
- Use of a fully digital DAQ system to allow processing of very high count rates, to minimize the dead time, optimize the time resolution of all the different detector types, and to make the signal processing simpler. Such a system can be run in either triggerless or triggered mode. Triggerless mode presents the advantage of a total freedom in the event reconstruction off-line.

Since the spectrometer is intended to be used with neutron beams no detectors are placed at forward angles (<20 degrees) to allow neutrons to escape the array without touching a detector – and potentially damage it. The $LaBr_3$ detectors are preferentially placed as two forward rings since they are much more resilient to scattered neutrons. A brief description of the ν-Balldetectors will be given in the following. More details can be found in [12].

2.0.1 The high purity germanium detectors

High purity germanium detectors were of two types: coaxial germanium crystals and clover detectors.

Ten coaxial germanium detectors, named "Phase I", were placed in a ring at backward angles (133.5° with respect to the beam axis) and 16 cm from the center of the sphere to the BGO shield entrance window. That sets the crystal at 18 cm from the center. They were rented to the FRENCH/UK loan pool [13] for the duration of the campaign.

The clover germanium detectors were borrowed from the GAMMAPOOL [8] for the duration of the campaign and are placed in two rings of twelve detectors each around 90° (75.5° and 104.5° with respect to the beam axis) to take advantage of their granularity in case of Doppler corrections. The distance from the BGO shield entry to the center was 17 cm which places the crystals at 24 cm from the center. Signals from germanium crystals and BGO crystals were directly digitized without preamplification. Post treatment of the data set will allow the use of the BGO signal for calorimetry purposes at the same time as for Compton rejection. A total photopeak efficiency of 4% and an energy resolution of 2.8 keV was measured at 1.33 MeV.

2.0.2 The $LaBr_3$:Ce detectors

The fast-timing part of the array was constituted of twenty $LaBr_3$:Ce crystals. These detectors were provided by the FATIMA [14] and UKNDN [15] collaborations. The remaining ten percent of available solid angle in the sphere permitted the packing of 20 of these detectors in two rings of ten detectors each. The first ring was at 46.5° with respect to the beam axis and 11 cm away from the sphere center. The second ring was at 34° with respect to the beam axis and 15 cm away from the sphere center. When coupled to the LICORNE directional neutrons source, the ν-Ball$LaBr_3$ second ring had to be moved further away from the center (~25 cm) in order to be outside the neutron cone. The reference layout of the array is presented in Fig. 1. A technical drawing from the IPNO design office of one hemisphere is presented in the right panel. Two rings of $LaBr_3$, two rings of clovers embedded in their BGO shields and a last ring of Phase I detectors in their shields are visible from left to right. The left panel shows a photograph of the array at the beginning of the experimental campaign before the mounting of the reaction chamber.

A separate configuration involving the coupling of ν-Ballwith four clusters of the PARIS [9–11] array was also performed. For this setup all the $LaBr_3$ detectors were replaced by 34 PARIS phoswitch detectors almost completely covering the available solid angle in the forward direction (~30%).

Globally, an energy resolution of 2.7% and an photopeak efficiency up to 1.5% (depending on the crystals positions) was measured at 662.7 keV.

2.1 The data acquisition system: FASTER

All the pre-amplifier or anode signals are sent to a fully digital acquisition system called FASTER (Fast Acquisition System for nuclEar Research) [16] developed at the Laboratoire de Physique Corpusculaire in Caen, France. It is used to measure relevant observables (deposited energy and time of each hit) extracted online from the traces. FASTER has been built to be modular and fully integrated to satisfy the needs of any type of experiment in nuclear physics, particularly those, which combine detector types with significantly different characteristics. FASTER is capable of on-line processing of traces, to sort them and group them into "events" (depending on each scientific goal) through a software event builder module, no matter what is the type of detector. At the same time, data can be visualized using the versatile ROOT Histogram Builder (RHB) software. Based on ROOT [17], it allows the implementation of online energy, time-of-flight (TOF) spectra or any kind of 1D-2D user-defined histograms. All the signals of the ν-Ballarray are sent to 184 (200 with the PARIS setup) independent electronics channels. These are distributed over 34 motherboards in three μTCA crates. The crates are synchronized using an external logic signal (called "t0") distributed by an MCH card located in a fourth crate. The maximum count rates, obtained during the campaign, in the individual detectors are: 10 kHz for the HPGe, 15 kHz for the $LaBr_3$ and 40 kHz for the BGO crystals without having any dead time, or showing a degradation of energy resolution.

2.2 The ν-Ballcampaign

The ν-Ballspectrometer was operational over a one year period from November 2017 to June 2018. Over 150 researchers participated in the experimental campaign and approximately 3200 hours of beam time were provided by the ALTO facility. A detailed description of the campaign progress is given in [18]. Eight experiments were performed, a short description of some of them is given in [19–21]. The variety of physics cases necessitated several transformations of the geometry during the course of the campaign.

After construction a commissioning experiment was performed using an ionisation chamber with a Frisch grid containing a ^{252}Cf spontaneous fission source of 3 kBq on a lead backing at the central cathode to stop one of the fission fragments. The chamber of 15 cm diameter was placed in the center of the array and the ν-Balldata acquisition triggered with the chamber cathode signal. The test was performed to validate the DAQ system at a low count rate with fission events of an average gamma multiplicity around 8. Data were acquired for around 50 hours.

Subsequently, a first set of experiments [19, 20] based on heavy-ion induced fusion-evaporation reactions were performed with all ν-Balldetectors placed in the most compact geometry possible – identical to the previously simulated GEANT4 setup. The trigger condition was multiplicity of two γ rays detected in any $LaBr_3$ module (or one Ge module and one $LaBr_3$) within 2 μs. Clover modules had a local trigger on the card, to provide a multiplicity of one, even in the event of more than one clover crystal firing.

Later experiments required a change in geometry to couple the array with the LICORNE neutron source [21]. The neutron source produces naturally directional neutrons in a cone that depends on the primary beam energy. It is important to remind that the neutron beam can be pulsed with a pulse width of 2 ns and a period of 400 ns.

The presence of the LICORNE chamber and the heavy metal collimator, that shield the gas cell, causes a partial shadowing effect on the backward ring of coaxial germanium detectors, thus slightly lowering the total efficiency of the array. A photograph of the nuball/LICORNE coupling is presented on the right of Fig. 2. The black detectors (conical $LaBr_3$ crystals) were retracted to a distance of 10 cm and the grey detectors (cylindrical $LaBr_3$ crystals) were retracted to a distance of 15 cm in order to make sure the source neutrons, produced in a cone of around 20 degrees opening angle, can escape the spectrometer. The experiments using a neutron beam to study the nuclear structure of very neutron-rich fission fragments and the fission mechanism itself were performed over seven weeks of beam time. The DAQ was ran in triggerless mode, and due to the very high count rates and network bandwidth limitations, data were sent to disk at maximum rates of around 70 MB/second, or 3 TB per day. The primary beam intensity was set to obtain counting rates in the individual crystals of around 10 kHz in the HPGe, 40 kHz in the BGO and 15 kHz in the $LaBr_3$. The total amount of data collected during the entire ν-Ballcampaign was very large (~0.2 PB), a significant proportion of which was transferred to the IN2P3 data-center [22].

The last experiment of the campaign replaced all the FATIMA $LaBr_3$ with PARIS phoswitches. A photograph of the coupling is presented in the right panel of Fig. 2. The 34 phoswitches were placed in a cylindrical cluster arrangement at a distance of 20 cm from

Figure 2. The left panel shows a photograph of the experimental setup of ν-Ballcoupled to the PARIS array. The right panel shows the ν-Ballarray coupled to the LICORNE neutron production chamber with a thick target of natural thorium placed at the center of the array.

the center of the array to maximize solid angle coverage. This addition provided a significant increase in the total efficiency at energies greater than 4 MeV. A further 14 FASTER CARAS channels were added to the system, making a total of 200 independent digital channels. This experiment was also run in triggerless mode to allow complete offline characterization of the cross-talk/scattering of γ-rays in the PARIS clusters.

3 Fission studies with ν-Ball

As mentioned in the previous section, a part of the beam time of the ν-Ballcampaign has been dedicated to the coupling with the LICORNE neutron source. It included three experimental projects on fission process studies: one dedicated to the search of fission isomer (experiment labeled: N-SI-103) and spectroscopy in the second potential well, the other two dedicated to study of fission fragments produced by the fast neutron induced fission of $^{238}_{92}$U and $^{232}_{90}$Th (experiment labeled: N-SI-109).

3.1 Description of fission experiment with ν-Ball

Experiment N-SI-103 used a cylindric aluminum tube in which five pellets of depleted uranium (isotopic purity of 99.7914%) were placed. The tube was 8 cm long and 3.1 cm diameter. The total mass of $^{238}_{92}$U was 81 grams and the effective density was 1.342 g/cm^3. This target was also used for the N-SI-109 experiment.

The $^{232}_{90}$Th target was made of a 8 cm long aluminum cone with an entry diameter of 2.5 cm and an exit diameter of 5.5 cm. Such a geometrical shape ensured a maximum overlap with the neutron spatial distribution out of the LICORNE gas cell. 7 pellets of natural thorium were placed to achieve a total mass of 129.7 g with an effective density of 0.984 g/cm^3. This target had an intrinsic activity of 5 MBq.

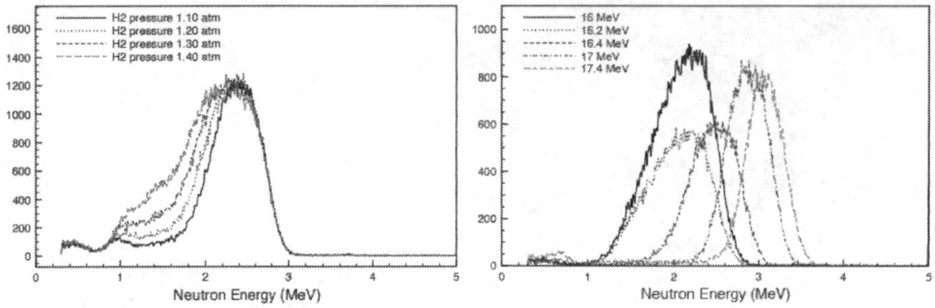

Figure 3. On the left: neutron energy spectra measured as a function of the H_2 pressure in the gas cell. On the right: neutron energy spectra measured as a function of the incident lithium beam energy.

The LICORNE beam was tuned to maximize the neutron flux in the target and minimize the cone opening to avoid hitting forward detectors. Figure 3 shows how the neutron spectrum can vary with two of the LICORNE parameters: incident lithium beam energy (on the right), and H_2 target density (on the left). For the two measurement a lithium beam energy of 16.75 MeV has been chosen.

3.2 Fission trigger for the ^{232}Th(n,f) experiment

While the beam was on target, several processes took place:

- ^{232}Th fission, with a calculated rate of 20 kHz.
- The p(^{7}Li,^{7}Li*)p reaction that produces a 478 keV γ-ray; An additional lead shielding has been designed to attenuate this γ source by a factor 15 . In addition as the γ-rays are emitted in flight, Doppler shift and broadening can be observed in the γ energy spectrum creating a structure between 470–515 keV that prevent any coincidence with γ-rays in this energy region. The expected reaction rate is in the order of 1 MHz with a 4π emission.
- With an specific activity of 4.09 kBq/g, this target achieved an activity of 5.2 MBq. Such an activity means a large probability of random coincidences even with the 2 ns narrow beam pulse and on a wide energy range.

After the first step of the data analysis: energy calibration, gain matching, time alignement (detailed in [12]), a first total projection of prompt $\gamma-\gamma$ matrix can be performed. The result is shown in Fig. 4. The projection of the $\gamma-\gamma$ matrix contains many γ lines produced by γ-rays from the parasitic processes detailed earlier. To minimize contamination with unwanted γ-rays, several trigger condition have been tested. All of them rely on the fact that fission process emits a larger number of γ-rays than the parasitic ones. For instance, a trigger considering prompt multiplicity (in a time window of 50 ns around the beam pulse) up to 6 have been considered. If the matrices are cleaned up of unwanted γ-lines it comes with a lower statistics. To try to compensate for this loss a new approach has been considered. It relies on a multivariate analysis using neural network architecture.

3.2.1 Neural networks

Neural networks (NN) is an algorithm capable of correlating a list of input values to a given list of output values. This correlation is created statistically in what is named the training process. The structure is composed of N neurons, each of them represent a smooth step function that for an input value between 0 and 1 returns a value in the same range. The slope

 https://doi.org/10.1051/epjconf/202125600008

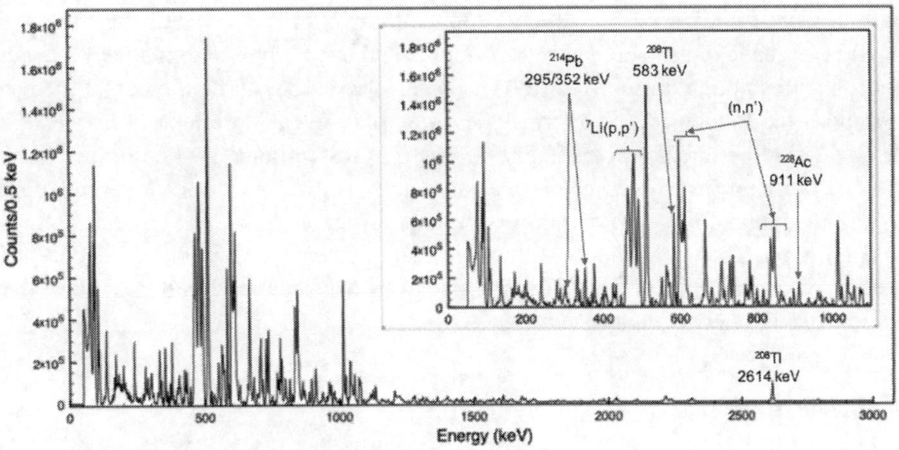

Figure 4. Total projection of a prompt-prompt $\gamma - \gamma$ matrix on the energy range 50–3000 keV. The subfigure represent a zoom on the low energy region.

of this step function and the normalisation factor at the input of the neuron is chosen during the training sequence. In other words, a neural network is capable of performing the selection of a subspace in an N dimension hyperspace with N, being the number of inputs which is very similar to a graphical cut on a 2D plot.

Neural ntworks are a branch of artificial intelligence that knows a rebirth of interest in science as it is an efficient way to get the results of complicated calculations with simple algorithms. It is now commonly used for instrumentations and physics application [23–25]. The design of a neural network is made of three major steps: understanding the useful input parameters, defining the architecture, the training procedure. Then, a neural network quality is characterized by its generalization capability. That is to say to give the correct output for a new dataset.

3.2.2 Determination of input data & NN architecture

The aim of the neural network is to discriminate fission from other nuclear reactions that occur at a much higher rate. The input parameters used for the NN will be measured using the ν-Ball spectrometer and are supposed to get distinguishable values depending on the detected reaction. At the moment of its conception, the array was known to have calorimetry capabilities, high resolution energy measurement and fast timing properties. The coupling of the spectrometer with the LICORNE neutron source made the use of pulsed neutron beams possible.

As a consequence, the first pre-treatment of the data will require that all the hits are grouped in an "event" that corresponds to a 400 ns time window around a beam pulse. More precisely the time window starts 60 ns before and end 340 ns after a beam pulse. From this set of hits several meta-data can be calculated such as: total and average energy (prompt or delayed w.r.t. the beam pulse), multiplicity (prompt or delayed), and standard deviation of these meta-data, etc. Up to 16 variables were identified. Some NN also use non-linear combination of inputs to increase the NN predictive power. In this first approach, this possibility was not considered.

The wanted output is a decision level: a real number between 0 and 1 that corresponds to the percentage of chances to have a fission. Later, depending on the response of the neural network a decision threshold can be chosen to optimize the discrimination.

With a defined number of inputs and outputs, the NN boundary conditions are defined: one input layer made of as many neurons as inputs and one output layer made of one neuron. In between some hidden layers need to be created. Depending on the layout of the neurons that compose the hidden layers, more or less complex linear correlations between inputs can be performed. There a unfortunately no precise prescriptions on the number of hidden layers of hidden neurons. Nonetheless, one can figure out that one hidden layer will correlate linearly all the inputs to a given output, two hidden layers will correlate quadratically all the outputs and so on. This mean that for our application two or three hidden layers consists of a reasonable maximum. The number of hidden neurons has been modified to achieve the best NN response after the training procedure.

3.2.3 Training Procedure

After a given NN layout has been chosen, all the normalisation factors and inputs weights for every neurons must be calculated. The ROOT software [17] provide a library named TMVAnalysis. This package has been created to be capable to quickly define a neural network structure and perform its training from lists of inputs taken from a ROOT file.

The next step is to build several data sets that will serve to train (fit) the neural network and to test its response. As mentioned earlier, there are three main processes that will be considered: fission, natural thorium target intrinsic activity and (p,p') reaction in the gas cell. Experimentally, several runs were acquired:

- During the commissioning of the spectrometer, an ionization chamber containing a ^{252}Cf spontaneous fission source was used. Considering the fission chamber anode signal as a "beam pulse", 400 ns wide events were reconstructed containing only fission. To these events, an output value of 1 was assigned.
- During the ^{232}Th(n,f) irradiation measurement, several calibration run were performed. During these runs, the RF signal was not switched off. Again, the same 400 ns wide events were reconstructed containing only target intrinsic activity events. To these events, an output value of 0 was assigned.
- During the ^{238}U(n,f) irradiation measurement, a run with the neutron beam but no target was performed. Then, the reconstructed 400 ns wide event contain mainly (p,p') reactions in the gas cell. To these events, an output value of 0 was assigned.

From these, two files containing meta-data of each event type and the expected output value were created. One file was used for training (named "Training Set") and the other one for checking the neural network response (named "Test Set"). In the training process, several file sizes were chosen. As the procedure is very similar to a fit, the initial naive step was to take a large number of training events. However, during the training procedure, it was shown to be not necessary.

The TMVAnalysis library training procedure consists of two steps. In the first one, the program will read all the inputs from the training set to determine normalization factors so that the input value cannot exceed 1. In a second step, it will fit all the weights of all the neurons on the test set. A user defined proportion of the test set can be used as a verification that the fit is good enough. If not, the neuron weights are reseted and the program tries a new fit.

At the end of the training procedure, the neural network is tested on a test set. The neural network response to new input data is calculated and the response is then compared with the known answer. The discrepancy between the response and the expected value serves as a parameter to evaluate a NN quality.

3.2.4 Neural network quality

If the observed discrepancy characterize the quality of the neural network to a test set, it might not be easy to quantify the generalization capability, *i.e.* getting the correct answer for an unknown set. Thus, for a given threshold on the output the rate of true positive *vs* false positive is a statistical information that can characterize more precisely the quality of a neural network generalization capability. It is called the receiver operating characteristic curve (or "ROC" curve) – see Fig. 5. In an ideal situation, it would be represented by a step function that takes the value 1 for any false positive rate value. In Fig. 5, continuous lines represent

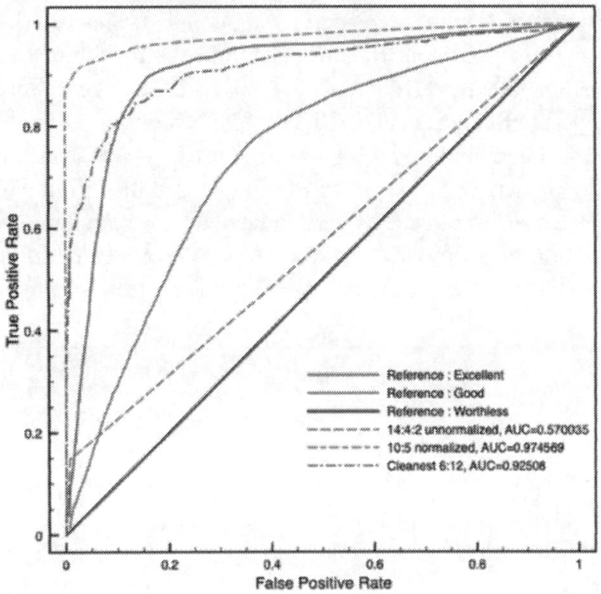

Figure 5. Receiver Operating Characteristic (ROC) curve for various neural network architecture: continuous lines represent limit cases; dashed lines represent actual network ROC curves.

prescriptions to evaluate the quality of a neural network using the ROC curve evaluator. The diagonal one represents the worst case scenario for which any decision threshold will give a 50% chance of mistake. If the NN has a ROC curve close to this one, then it is pretty much useless. The green and red curves represent a good and excellent cases, respectively. If the ROC curve is above the red one, then the neural network might be too good. In our situation, too good means that the neural network has perfectly learnt the features of the spontaneous fission of ^{252}Cf detected by a fission chamber. However it fails to generalize the features to other fissioning systems that are not detected in a fission chamber. In the domain of artificial intelligence this corresponds to overtraining. In Fig. 5, the curve labeled "10:5 normalized" is a perfect example with a success rate of 90–100% to recognize fission in the test set. When applied to the thorium data set, no fission fragment specific γ peaks are visible in a $\gamma - \gamma$ matrix.

During the training process:

- More than fifty architectures were tested.
- Different training sets were tested. The size was changed from a few thousand events to several millions.

- The representation of each process in the training set was also studied: equally likely, same ratio as reaction rate ratios, ...
- The expected output value was also modified to include the likeliness of processes. For example, one can assume that the (p,p') reaction is more similar to intrinsic decay in the target than to fission when comparing the input data: the expected response of the NN was 0 for both intrinsic activity and (p,p') reaction.

At the end, one neural network architecture has been selected. It is using six metadata as inputs: the prompt spatial distribution of γ-rays, the 400 ns average spatial distribution, the event/prompt multiplicity and the event/prompt total energy. Its ROC curve is named "Cleanest 6:12" in Fig. 5. It is close to the reference for excellent neural networks. This architecture was then used to perform the fission trigger decision in the thorium experimental data. An analysis code was written to extract from real data pulses the necessary inputs. Then, a threshold (corresponding to a confidence level of 70%) on the NN output was set. From the selected data, a prompt-prompt (in coincidence with a beam pulse) $\gamma - \gamma$ matrix was constructed. The total projection of this matrix is presented in Fig. 6 as a red histogram. In the sub-view, the $4^+ \rightarrow 2^+$ and $2^+ \rightarrow 0^+$ transitions in ^{140}Xe are indicated. For comparison purposes, the blue histogram is a projection of a similar matrix based on a prompt multiplicity filter. The two projections have been normalized. In matrix projection resulting of a neural network based

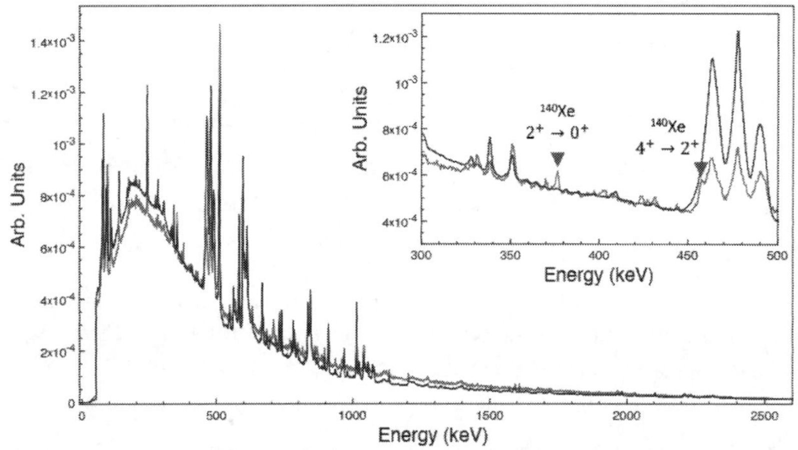

Figure 6. Total projection of two prompt-prompt $\gamma - \gamma$ matrices. The red one is obtained using a neural network based trigger. The blue one is obtained using a multiplicity filter of at least three HPGe crystal fired (after add-back and Compton suppression) in coincidence with the beam pulse. The sub-view is a zoom on the low energy region. The $4^+ \rightarrow 2^+$ and $2^+ \rightarrow 0^+$ transition of the most produced fission fragment (^{140}Xe) are shown.

decision, the main γ peaks produced by the γ-rays emitted by the most produced fission fragment are clearly visible. It is an achievement in itself as these peaks are generally in an energy range dominated by parasitic reactions. Again, these fission fragment induced structures are not visible on the matrix projection built with a multiplicity filter. The rejection rate of the neural network algorithm seems to be reasonable as the measured fission rate (measured with the ^{140}Xe $2^+ \rightarrow 0^+$ transition) is about the same order of magnitude as the calculated one. However, this success has to be mitigated. Indeed, the measured fission rate greatly depends on the considered fission fragment and it might vary from the expected value to a fifth of the calculated rate. Added to this, construction of cubes and matrices based on the selected data leaves a very small amount of data, which makes the identification of weakly populated transition almost impossible. In that sense a more classical multiplicity trigger is still a robust way to address this type of data.

To conclude, the high selectivity of the NN algorithm has been demonstrated. This first try to apply that type of triggering technique highlighted the pro and the cons of such procedure. First, during experiment, no data acquisition dedicated to the construction of training set has been performed just before the actual experiment. This means that the geometry, the number of channels available, or more globally the response function of the array has changed between the training set and the experimental set we apply the neural network to. Second, the fission data set was based on a different fissioning system: ^{252}Cf instead of ^{232}Th. If globally, the fission process ought to generate high γ multiplicity events with a high total energy compare to the parasitic processes, one may argue that ^{232}Th might get closer to multiplicity and energy released by the intrinsic activity of the target [26], making the discrimination even more difficult. Last, the library used to build the neural network structure was using basic functions, a more evolved one based on python programming language [27] might provide better results.

4 Conclusion

In this paper, a brief overview of the ν-Ballcampaign has been given. The ν-Ballarray, built in a similar way to EUROBALL, EXILL, is one of the first to combine high energy resolution detectors with high time resolution crystals. During the campaign, a significant beam time has been dedicated to induce fission on ^{238}U and ^{232}Th with fast neutrons produced by the ALTO LICORNE source. The high activity of the natural thorium target, made the fission event reconstruction difficult. Thus, a new technique based on neural network has been studied. It has been proven to be successful. However, several improvements can be performed and should be implemented in the future. If given the opportunity, the construction of a real fission dataset of the fissioning system with exactly the same response function of the array is a must have. In the context of a ν-Ball2 campaign it will have to be prepared.

References

[1] U.F.E. Collaboration, Nucl. Phys. A **520**, c657 (1990)
[2] J. Eberth et al., Prog. Part. Nucl. Phys. **28**, 495 (1992)
[3] F.A. Beck, Prog. Part. Nucl. Phys. **28**, 443 (1992)
[4] I.Y. Lee, Nucl. Phys. A **520**, c641 (1990)
[5] C. Sotty, AIP Conf. Proc. **1852**, 080009 (2017)
[6] V. Werner et al., J. Phys. Conf. Ser. **366**, 012048 (2012)
[7] D. Bucurescu et al., Nucl. Instr. and Meth. in Phys. Res. A **837**, 1 (2016)
[8] http://gammapool.lnl.infn.it/index.htm
[9] A. Maj et al., Acta Phys. Pol. B **40**, 565 (2009)
[10] M. Zieblinski et al., Act. Phys. Pol. B **44**, 651 (2013)
[11] C. Ghosh et al., J. Instrum. **11**, P05023 (2016)
[12] M. Lebois, N. Jovancevic, D. Thisse, R. Canavan, D. Etasse, M. Rudigier, J. Wilson, Nucl. Instrum. Methods Phys. Res. A **960**, 163580 (2020)
[13] http://ipnwww.in2p3.fr/GePool/poolRules.html
[14] O.J. Roberts et al., Nucl. Instrum. Methods A **748**, 91 (2014)
[15] http://www.ukndn.ac.uk

[16] http://faster.in2p3.fr
[17] R. Brun, F. Rademakers, Nucl. Instrum. Methods Phys. Res. A **389**, 81 (1997)
[18] M. Lebois et al., Acta Phys. Pol. B **50** (2019)
[19] M. Rudigier et al., Acta Phys. Pol. B **50** (2019)
[20] S. Leoni et al., Acta Phys. Pol. B **50** (2019)
[21] N. Jovancevic et al., Acta Phys. Pol. B **50** (2019)
[22] https://cc.in2p3.fr/en/
[23] E. Ronchi et al., Nucl. Instrum. Methods A **610**, 534 (2009)
[24] Z. Szadkowski et al., Progress in Electromagnetics Reseach Symposium (2017)
[25] L. Gupta et al., Pattern Recog. **25**, 413 (1992)
[26] J.W. Wilson et al., THEORY-5 proceeding to be published (2020)
[27] https://docs.python.org/3/reference/index.html

Anisotropy in fission fragment and prompt neutron angular distributions

A.E. Lovell[1,*], *P.* Talou[1], *I.* Stetcu[1], and *K.J.* Kelly[1]

[1]Los Alamos National Laboratory, Los Alamos, New Mexico, 87545, USA

Abstract. Several physics mechanisms can lead to the deviation from an isotropic angular distribution for both fission fragments and the neutrons that are emitted during the fission event. Two of these effects have recently been implemented into CGMF, the Monte Carlo fission event generator developed at Los Alamos National Laboratory: angular distribution sampling for fission fragments and pre-equilibrium neutrons (those emitted before the compound nucleus forms). Using these new developments, we show that the anisotropy of the neutrons reflects the anisotropy of the fission fragments, in particular as the outgoing energy of neutrons increases. Correlations between the fission fragment and neutron anisotropies could be used to extract the fission fragment anisotropy from the neutron angular distributions.

1 Introduction

Accurately describing the fission process is important for many applications from the formation of the compound nucleus to the emission of prompt and β-delayed neutrons and γ rays. A few codes are available to calculate the emission of neutrons and γ rays during fission events, either on an event-by-event basis or through direct calculations of the prompt neutron and γ-ray observables [1–6]. For event-by-event models used to study correlations between fission observables, it is also important to be able to describe the angular distributions of the fission fragments and emitted particles. The opening angle between pairs of neutrons or neutrons and fission fragments can be calculated and compared with experimental data, e.g. [7] and references therein.

In this work, we focus on the angular distributions of the fission fragments and prompt neutrons. There are several mechanisms that could cause these angular distributions to be anisotropic. For neutron-induced fission, momentum conservation between the incident neutron and the fissile target boosts the fragments and emitted particles along the beam axis in the forward direction. The fission fragments are also emitted in a preferential direction with respect to the beam axis [8] that depends on the incident neutron energy and is due to the fission transition states that are populated at scission. The angular distribution of the pre-equilibrium neutrons (emitted before a compound nucleus is formed) is forwarded peaked [9], compared to the other isotropically emitted pre-fission neutrons. In addition, neutrons emitted from the fission fragments could have a non-isotropic angular distribution from the spin of the fission fragments or could be emitted from partially accelerated fragments – instead of fully accelerated fragments – which would cause these angular distributions to be anisotropic in the lab

*e-mail: lovell@lanl.gov

frame. Attempts have been made to describe the emission of neutrons from accelerating fission fragments, e.g. [10], and models such as the Time-Dependent Superfluid Local Density Approximation (TDSLDA) [11] could provide useful insights into this question. While work is ongoing to experimentally investigate whether neutrons are emitted from fully accelerated fragments or not, most often, the discrepancy between modeled and experimental neutron angular distributions are attributed to scission neutrons–although the percentage of neutrons emitted in this manner is still somewhat of a controversy [12–16]. In this work, we focus on the effects of kinematics, fission fragment anisotropies, and pre-equilibrium neutron angular distributions. While the kinematics of the beam have always been taken into account, the latter two effects have recently been implemented into the fission event generator developed at Los Alamos National Laboratory (LANL), CGMF [1, 2], and here, we show the effects on the fission fragment and neutron anisotropies when these models are included.

2 Theory

In this section, we briefly describe the theories behind the fission fragment anisotropies and pre-equilibrium neutron emission. We will also give some details about the CGMF code.

2.1 Fission fragment angular distributions

Angular anisotropies of fission fragments as a function of incident neutron energy have been measured for several fissile targets [17–27] and have been shown to change as the incident energy increases. It was postulated [8] that this preferential direction of fragment emission comes from the transition states in the compound nucleus that are populated at the outer saddle point.

The angular anisotropy is defined as a ratio of the cross section at 0° to the cross section at 90°, $W(0^\circ)/W(90^\circ)$. The angular distributions can be described by [28],

$$W(\theta) = \sum_{s(J,K,M)} P_s(J,K,M)\frac{2J+1}{2}\left|d^J_{M,K}(\theta)\right|, \tag{1}$$

where $P_s(J,K,M)$ is the population of each state, s, identified by the total angular momentum of the compound nucleus, J, its projection onto the fission axis, K, and its projection onto the beam axis, M. The angular dependence is encoded in the Wigner d-matrices, $d^J_{M,K}(\theta)$.

2.2 Pre-equilibrium neutron angular distributions

At high enough incident neutron energies, pre-fission neutrons can be emitted before the compound nucleus forms, i.e. pre-equilibrium neutrons. The energy spectrum and angular distribution of these neutrons differ from the other pre-fission neutrons that are emitted from a compound nucleus. These neutrons can be described by the quantum mechanical theory of Feshbach, Kerman, and Koonin (FKK) [29], in which the angular distribution can be separated into a multi-step direct and multi-step compound part. The pre-equilibrium neutrons are described by the multi-step direct piece of this formulation and are similar to inelastically scattered neutrons.

Recently, the angular distribution of pre-equilibrium neutrons was extracted experimentally for the first time for ^{239}Pu(n,f) [30]. The pre-equilibrium channel opens around $E_{\rm inc}$ = 12 MeV, and the neutron angular distributions are significantly forward peaked. These angular distributions are well-described by the FKK theory and can also be reproduced by Kalbach systematics [31].

2.3 CGMF

CGMF is a Monte Carlo Hauser-Feshbach code developed at LANL to describe the evaporation of prompt neutrons and γ rays emitted from fission fragments [1, 2]. The required inputs are scission-fragment yields in mass, charge, total kinetic energy, spin, and parity, $Y(A, Z, TKE, J, \pi)$. The mass yields, $Y(A)$, are three-Gaussian fits to experimental data [32, 33], and the charge yields, $Y(Z|A)$, are taken from Wahl systematics [34]. The TKE distribution for each mass, $Y(TKE|A)$, is also taken to be Gaussian. The spin of each fragment is sampled from $Y(J) = (J+1)\exp(-\alpha J(J+1)/2B^2)$ where B^2 depends on the nuclear temperature and moment of inertia and α is a global scaling parameter [35]. The parity is evenly sampled, either +1 or −1.

Once the initial conditions are sampled, the total excitation energy, TXE, is determined from the Q-value, TXE $= Q -$ TKE. The excitation energy is then shared between the two fragments based on a ratio, R_T, which is fitted as a function of mass to reproduce the average neutron multiplicity, $\overline{\nu}(A)$. From the excitation energy of each fragment, neutrons and γ rays are evaporated probabilistically through the Hauser-Feshbach statistical theory. The neutron probabilities are calculated through transmission coefficients using an optical potential parameterization (Koning-Delaroche, by default), and the γ-ray probabilities are determined from the γ-ray strength function for E1, M1, and E2 transitions. The fissions are calculated on an event-by-event basis, and for each event, the initial conditions of the fission fragments are recorded along with the momenta and energies of the fragments, neutrons, and γ rays. In this way, both average and correlated quantities can be constructed.

3 Fitting procedure

In this section, we describe the process for fitting and sampling the fission fragment anisotropies and pre-equilibrium angular distributions, along with describing how the fragment and neutron anisotropies are extracted from the CGMF calculations.

3.1 Fission fragments

For most of the fissile targets considered in CGMF, there are several sets of experimental data across the incident neutron energy range of interest, from thermal to 20 MeV. For each target, we fit the available data using cubic splines. One spline fit compared to experimental data is shown in Fig. 1 for ^{239}Pu(n, f). There is good coverage of the ^{239}Pu(n, f) data across the whole energy range, and with only a few knots, we are able to reproduce the fluctuations in the fission fragment anisotropy.

The angular dependence in cosθ of the light fragment is sampled from the distribution,

$$P(\cos\theta) \propto 1 + \left(\frac{W(0°)}{W(90°)} - 1\right)\cos^2\theta. \quad (2)$$

When $W(0°)/W(90°) = 1$, cosθ is sampled isotropically. In all cases, ϕ is sampled isotropically. The direction of the heavy fragment is constructed through momentum conservation in the center-of-mass frame of the compound nucleus.

Relying on a cubic spline fit does limit our inclusion of the fission fragment anisotropy to isotopes where data have been measured over a wide range of incident energies. For ^{241}Pu(n, f), data are only available up to $E_{\rm inc} = 8$ MeV. In this case, the data over this range have the same shape and magnitude as the anisotropy for ^{239}Pu(n, f) so we are able to use the same fit as a function of incident neutron energy. Using a spline fit implicitly accounts for

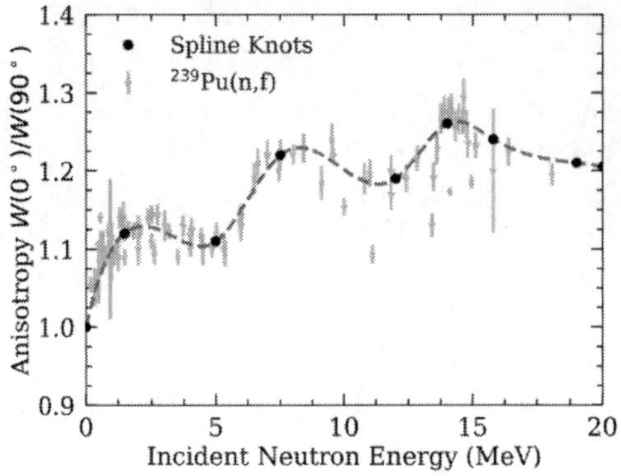

Figure 1. Fission fragment anisotropy as a function of incident neutron energy for ^{239}Pu(n, f). Data is shown as the green points, spline knots as black filled circles, and the cubic spline fit as the dashed dark green line.

multi-chance fission about the neutron separation energy which makes this approach only an approximation. In principle, Eq. (1) could be used to construct these angular distributions, the weights $P_s(J, K, M)$ are unknown and are often fitted to the measured angular distributions, instead of predicted directly from theory.

3.2 Pre-equilibrium neutrons

Angular distributions for the pre-equilibrium neutrons were calculated for ^{239}Pu using the quantum mechanical theory of Feshbach, Kerman, and Koonin [29] over a range of incident neutron energies and target excitation energies [36]. This calculation gives an approximate shape for the angular distributions due to the limited model space. To smooth out any irregularities and to have a parameterization across all incident and excitation energies, the angular distributions were fitted using Kalbach systematics [31],

$$\frac{d\sigma}{d\Omega} = a(E_{\text{inc}}, E^*)\sinh(\cos\theta) + b(E_{\text{inc}}, E^*), \tag{3}$$

where E_{inc} is the incident neutron energy and E^* is the excitation energy of the target system. The parameters a and b are taken to be polynomial in E_{inc} and E^*. Although these calculations were performed for ^{239}Pu, the same fits were used for all targets within `CGMF` as the addition or removal of a few nucleons does not significantly change the calculated angular distributions.

3.3 Extracting the angular anisotropy

To extract the angular anisotropy from the `CGMF` calculations, we bin the calculated $\cos\theta$ values then fit these angular distributions to a sum of Legendre polynomials,

$$W(\theta) = p_0 L_0(\cos\theta) + p_1 L_1(\cos\theta) + p_2 L_2(\cos\theta), \tag{4}$$

where p_0, p_1, and p_2 are the fitting parameters weighing each of the Legendre polynomials. Since the fission fragment angles are sampled from a $\cos^2\theta$ distribution, we should not need any higher order terms to describe these angular distributions. The $L_1(\cos\theta)$ linear term is included to take into account the boost of the fragments from the center-of-mass frame to the laboratory frame. The three Legendre polynomial parameterization of Eq. (4) is used to fit the angular distribution of the neutrons as well.

We use this fitting procedure instead of directly comparing the number of counts at 0° and 90° to mitigate effects of low statistics on the mean values of the anisotropy, in particular when we begin to consider the angular distributions of neutrons with an increasing energy threshold, E_{thres}. This fitting procedure also allows us to extract uncertainties on the resulting anisotropy values. The covariance between the best-fit parameters of Eq. (4) can be constructed. A distribution of parameter sets can then be calculated by sampling a multi-variate Gaussian distribution centered around the best-fit parameter values with a variance given by the constructed covariance matrix. These parameter sets can then each be used to calculate the anisotropy from Eq. (4). Two thousand anisotropy values are calculated for each energy of interest, and 95% confidence bands are constructed by removing the most outlying 5% of the calculations.

4 Results

We can first look at the changes to the fission fragment and neutron anisotropies when kinematics, fission fragment anisotropy, and pre-equilibrium angular distributions are included into `CGMF`. The effects of kinematics had already been implemented into `CGMF`. The anisotropy of the fission fragments, neutrons emitted from the fission fragments, and all neutrons (from the fission fragments and the pre-fission process) from just the recoil is shown in Fig. 2a. Here, the anisotropy of the fission fragments is less than 5% at E_{inc} = 20 MeV, and as the neutrons are emitted either isotropically from the fission fragments or from the target nucleus, the anisotropy of the neutrons is lower. Figure 2b shows the anisotropies after the fission fragment anisotropy is taken into account, and c shows the anisotropies when the forward-peaked pre-equilibrium angular distributions are sampled. In Fig. 2b, there is little difference between the anisotropies of the neutrons when pre-fission neutrons are or are not considered, except due to the slight differences in the kinematics of the compound after a pre-fission neutrons is emitted. When the angular distribution of the pre-equilibrium neutrons is taken into account, there is a noticeable increase in the overall anisotropy of all neutrons.

In each case, the neutron anisotropy tracks the fission fragment anisotropy but is overall lower in magnitude, as the angular distributions are flatter compared to those of the fission fragments. Also, when the pre-equilibrium neutron angular distributions are sampled, we see very little change in the fission fragment anisotropy and the anisotropy of the neutrons emitted directly from the fission fragments. Above about E_{inc} = 12 MeV, the anisotropy increases when all of the neutrons are considered, where the pre-equilibrium neutrons are first emitted. Their forward-peaked angular distributions are clearly reflected in the resulting neutron anisotropy, even though they account for less than 20% of the emitted pre-fission neutrons.

As the outgoing neutron threshold energy, E_{thres}, increases, the neutron angular distributions – and thus the neutron anisotropy – tends toward that of the fission fragments. In Fig. 3, we show the neutron anisotropy as a function of incident neutron energy compared to the measured fission fragment anisotropy for a the neutrons emitted from the fission fragments and b when all emitted neutrons are taken into account. As the neutron threshold energy increases, the neutron anisotropy more closely tracks the fission fragment anisotropy, but the uncertainty on the extracted anisotropy also increases. This is due to statistics; the higher energy neutrons are more rare than the low-energy neutrons, and the fitting procedure of the Legendre polynomials becomes more uncertain.

From Fig. 3b, we also see that including the pre-fission neutrons in this analysis complicates the picture. At the highest incident energies, the neutron anisotropies are larger than

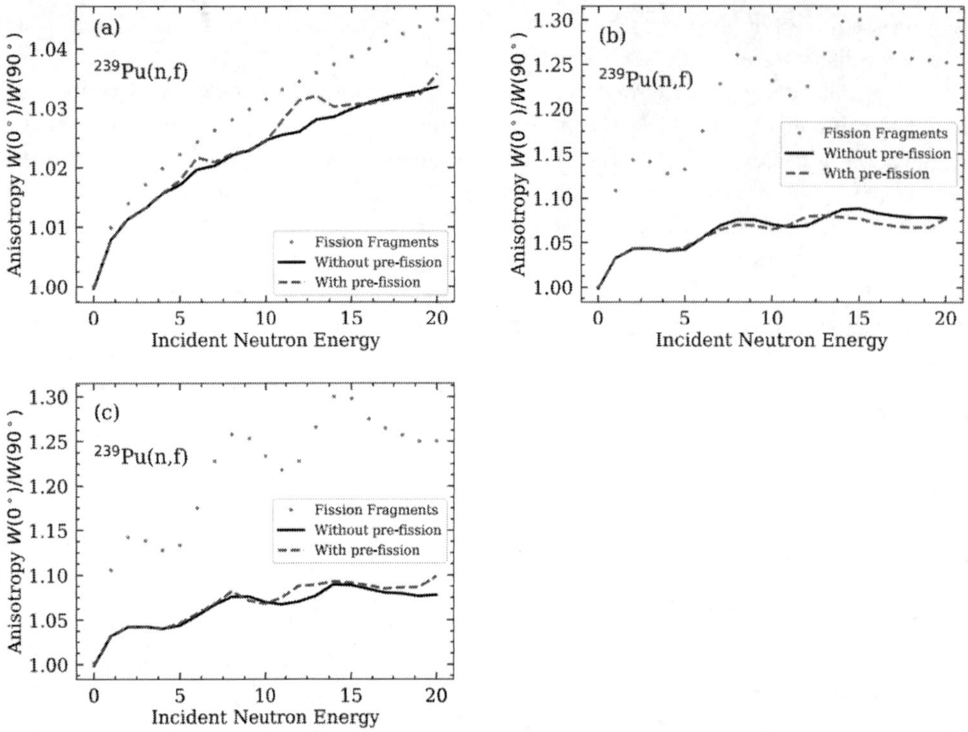

Figure 2. Comparison of the anisotropy for fission fragments, neutrons emitted only from the fragments, and all neutrons emitted during the fission reaction for ^{239}Pu(n, f) when (a) only recoil is included (fission fragments and pre-equilibrium neutrons are sampled isotropically), (b) recoil and the fission fragment anisotropy are included, and (c) recoil, fission fragment anisotropy, and pre-equilibrium neutron angular distribution sampling are included.

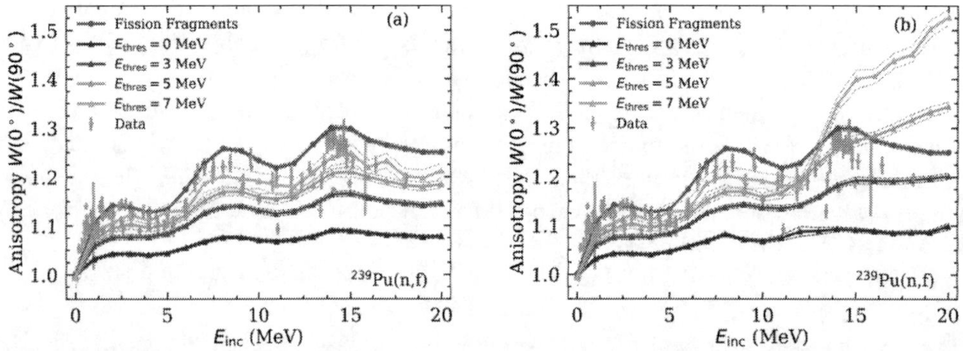

Figure 3. Neutron anisotropy as a function of incident neutron energy for $E_{\mathrm{thres}} = 0, 3, 5$, and 7 MeV (in indigo, blue, turquoise, and cyan, respectively) compared to experimentally measured fission fragment anisotropy (grey stars) for (a) neutrons emitted from the fission fragments and (b) all neutrons emitted during the fission event for ^{239}Pu(n, f). Also for comparison, the fission fragment anisotropy calculated from `CGMF` in dark grey.

the fission fragment anisotropies and become increasingly so as the threshold energy rises. However, as the threshold energy rises, the incident energy where the pre-equilibrium neutrons have an effect also increases. The increase in the anisotropy caused by the pre-fission neutrons is pushed to higher incident energies. Therefore, the incident energy to which the fission fragment anisotropy can be extracted from the neutron anisotropy increases if the neutron threshold energy can be increased.

There is, of course, a trade-off between the neutron threshold energy and the statistics. Fewer neutrons are emitted with higher outgoing energies. For codes such as `CGMF`, these challenges can be mitigated by running longer calculations. Collecting more statistics experimentally is a larger effort. However, for a given incident energy and threshold energy, a correction factor could be deduced from `CGMF` to convert the neutron anisotropy to the fission fragment anisotropy.

5 Conclusion

In this work, we have newly included two sources of anisotropy into the fission decay code `CGMF`: the angular anisotropy of the fission fragments coming from the transition states that are populated at the outer saddle point and of the pre-equilibrium neutrons. In addition to the angular anisotropies that are induced by the recoil of the target due to the incident neutron, which had already been taken into account in `CGMF`, these effects lead to clear angular anisotropies in both the fission fragments and the prompt neutrons that are emitted. These effects have now been taken into account in `CGMF` for neutron-induced fission on ^{233,234,235,238}U, ^{237}Np, and ^{239,241}Pu.

To include the fission fragment anisotropy, experimental data was fit with cubic splines and the fission fragment angles were sampled from a $\cos^2\theta$ distribution. Pre-equilibrium neutron angular distributions were calculated across a range of incident energies and excitation energies using the Feshbach, Kernan, and Koonin quantum mechanical model and fit globally to Kalbach systematics. The anisotropy of the prompt neutrons reflects the anisotropy of the fission fragments as a function of incident neutron energy, although the neutron anisotropy is significantly lower. By removing lower energy neutrons from the analysis, the magnitude of the neutron anisotropy comes closer to that of the fission fragments. However, these events occur less frequently, and the analysis begins to suffer from lack of statistics. Still, by utilizing the correlations between the fission fragment and neutron anisotropies seen in `CGMF`, the fission fragment anisotropy could be extracted from a measurement of the neutron anisotropy.

This work was performed under the auspice of the U.S. Department of Energy by Los Alamos National Laboratory under Contract 89233218CNA000001 and was supported by the Office of Defense Nuclear Nonproliferation Research & Development (DNN R&D), National Nuclear Security Administration, U.S. Department of Energy. We gratefully acknowledge the support of the U.S. Department of Energy through the LANL/LDRD Program and the Center for Non Linear Studies.

References

[1] P. Talou, I. Stetcu, P. Jaffke, M. Rising, A. Lovell, T. Kawano, Comput. Phys. Commun. **269**, 108087 (2021)

[2] B. Becker, P. Talou, T. Kawano, Y. Danon, I. Stetcu, Phys. Rev. C **87**, 014617 (2013)

[3] J. Verbeke, J. Randrup, R. Vogt, Computer Physics Communications **191**, 178 (2015)

[4] J. Verbeke, J. Randrup, R. Vogt, Computer Physics Communications **222**, 263 (2018)

[5] S. Okumura, T. Kawano, P. Talou, P. Jaffke, S. Chiba, J. Nucl. Sci. Tech. **55**, 1009 (2018)

[6] O. Litaize, O. Serot, L. Berge, The European Physical Journal A **51**, 177 (2015)

[7] P. Talou, R. Vogt, J. Randrup, M.E. Rising, S.A. Pozzi, J. Verbeke, M.T. Andrews, S.D. Clarke, P. Jaffke, M. Jandel *et al.*, The European Physical Journal A **54**, 9 (2018)

[8] A. Bohr, *On the Theory of Nuclear Fission*, in *International Conference on the Peaceful Uses of Atomic Energy*, edited by N.Y. United Nations, Vol. 2 (Geneva, 1956), p. 151

[9] E. Gadioli, P.E. Hodgson, *Pre-Equilibrium Nuclear Reactions* (Oxford Scientific Publications, 1992)

[10] A. Matsumoto, H. Taninaka, K. Hashimoto, T. Ohsawa, Journal of Nuclear Science and Technology **49**, 782 (2012), https://doi.org/10.1080/00223131.2012.677127

[11] A. Bulgac, *et al.*, in progress (2021)

[12] H.R. Bowman, S.G. Thompson, J.C.D. Milton, W.J. Swiatecki, Phys. Rev. **126**, 2120 (1962)

[13] H.R. Bowman, J.C.D. Milton, S.G. Thompson, W.J. Swiatecki, Phys. Rev. **129**, 2133 (1963)

[14] S.S. Kapoor, R. Ramanna, P.N.R. Rao, Phys. Rev. **131**, 283 (1963)

[15] K. Skarsvåg, K. Bergheim, Nuclear Physics **45**, 72 (1963)

[16] N. Carjan, M. Rizea, Phys. Rev. C **99**, 034613 (2019)

[17] A.S. Vorobyev, A.M. Gagarski, O.A. Shcherbakov, L.A. Vaishnene, A.L. Barabanov, JETP Letters **102**, 203 (2015)

[18] S. Ahmad, M.M. Islam, A.H. Khan, M. Khaliquzzaman, M. Husain, M.A. Rahman, Nuclear Science and Engineering **71**, 208 (1979)

[19] J.E. Brolley, W.C. Dickinson, R.L. Henkel, Phys. Rev. **99**, 159 (1955)

[20] J.E. Simmons, R.L. Henkel, Phys. Rev. **120**, 198 (1960)

[21] V. Geppert-Kleinrath, F. Tovesson, J.S. Barrett, N.S. Bowden, J. Bundgaard, R.J. Casperson, D.A. Cebra, T. Classen, M. Cunningham, D.L. Duke *et al.* (NIFFTE Collaboration 2), Phys. Rev. C **99**, 064619 (2019)

[22] R.L. Henkel, J.E. Brolley, Phys. Rev. **103**, 1292 (1956)

[23] E. Birgersson, A. Oberstedt, S. Oberstedt, F.J. Hambsch, Nuclear Physics **817**, 1 (2009)

[24] I. Ryzhov, M. Onegin, G. Tutin, J. Blomgren, N. Olsson, A. Prokofiev, P.U. Renberg, Nuclear Physics **760**, 19 (2005)

[25] F. Vivès, F.J. Hambsch, H. Bax, S. Oberstedt, Nuclear Physics **662**, 63 (2000)

[26] R.B. Leachman, L. Blumberg, Phys. Rev. **137**, B814 (1965)

[27] L. Blumberg, R.B. Leachman, Phys. Rev. **116**, 102 (1959)

[28] C. Wagemans, *The Nuclear Fission Process* (CRC Press, 1991)

[29] H. Feshbach, A. Kerman, S. Koonin, Annals of Physics **125**, 429 (1980)

[30] K.J. Kelly, T. Kawano, J.M. O'Donnell, J.A. Gomez, M. Devlin, D. Neudecker, P. Talou, A.E. Lovell, M.C. White, R.C. Haight *et al.*, Phys. Rev. Lett. **122**, 072503 (2019)

[31] C. Kalbach, Phys. Rev. C **37**, 2350 (1988)

[32] I. Stetcu, P. Talou, T. Kawano, EPJ Web Conf. **146**, 04026 (2017)

[33] I. Stetcu, P. Talou, T. Kawano, EPJ Web of Conferences **122**, 01012 (2016)

[34] A.C. Wahl, Tech. rep., Los Alamos Nat. Lab. LA-13928 (2002)

[35] I. Stetcu, P. Talou, T. Kawano, M. Jandel, Phys. Rev. C **90**, 024617 (2014)

[36] T. Kawano, private communication (2018)

Microscopic description of α-decay as super-asymmetric fission

Mihail Mirea[1,2,*]

[1]Horia Hulubei National Institute for Physics and Nuclear Engineering, P.O. Box MG-6, 077125 Bucharest-Magurele, Romania
[2]Academy of Romanian Scientists, Splaiul Independentei 54, 050094 Bucharest, Romania

Abstract. The fine structure of α-decay is treated with fission-like models. The single particle levels are calculated along a least action path connecting the ground state of the parent nucleus and the configuration of two spherical tangent nuclei. The probabilities to find different seniority-1 configurations are obtaining by solving the time-dependent pairing equations generalized by including the Landau-Zener effect and the Coriolis coupling. The theoretical results for the α-decay of ^{211}Po and ^{211}Bi are compared with experimental data showing a good agreement.

1 Introduction

The α emission is usually explained as the formation of a cluster on the surface of the daughter nucleus followed by a penetrability of an external barrier at an energy close to the Q-value [1–15]. The probability to form such a combination is given by a square overlap of the convolution of the wave functions of the parent nucleus and that of the system at the touching configuration. The preformation amplitude depends on the initial and final states, without taking into account intermediate configurations. Recent advances for a more realistic treatment were obtained within the quartetting wave function approach [16–19].

In fission-like theories [20–22] of α-decay, an extreme saturation of the nuclear matter is assumed [23]. Accordingly, the states of the nucleons should be strongly dependent on the boundaries of the many-body potential. When a preexisting α particle inside the parent nucleus is displaced towards the nuclear surface region, it perturbs all the microscopic states of the compound system. So, in a fission-like model, the rearrangement of the intrinsic states of the system during the formation of the combination consisting in both an α and a daughter nuclei should be taken into consideration.

In the formalism presented in the following, the rearrangement of the microscopic states during the emission of an α-particle is investigated dynamically. At scission, we are left with two mean fields, one corresponding to the α cluster and another pertaining to the daughter nucleus. The probabilities to have the system in different final configurations should be determined.

For this purpose, a system of microscopic differential equations is deduced. These equations are managed by the variational principle [24]. They represent a generalization of the

*e-mail: mirea@ifin.nipne.ro

time-dependent pairing equations by including two mechanisms: the Landau-Zener effect and the Coriolis coupling. Also, these equations provide information about the energy dissipated by the system. But in same time, we have an information concerning the distributions of intrinsic and collective angular momenta, and the momentary probabilities of different intrinsic configurations. The rearrangement in time of the nuclear orbitals during the disintegration process, starting from the initial compound nucleus and reaching the scission is investigated. As an application, the fine structure of the alpha-decay of an odd-A mass nucleus is treated as the simplest modality to test the theory. The dynamics of the α-decay process is revealed. From the least action principle, a superasymmetric fission path was obtained. The deformation energy is obtained in the framework of the macroscopic-microscopic model. The level scheme is calculated within the Woods-Saxon two center shell model. Consequently, the intensities of the transitions to the excited states of the daughter were evaluated theoretically. The experimental data were compared with the theoretical findings. A very good agreement was obtained. A mean value of the tunneling velocity of about 2×10^4 fm/fs was extracted.

2 Microscopic equations of motion

The coupled system of differential equations are obtained from the variational principle:

$$\delta L = \delta \left\langle \varphi_{IM} \left| H + H_R - i\tfrac{\partial}{\partial t} + H' \quad -\lambda |N_2\hat{N}_1 - N_1\hat{N}_2| \right| \varphi_{IM} \right\rangle \tag{1}$$

where H is the many-body Hamiltonian with pairing residual interaction, H_R is the centrifugal contribution, H' is a term simulating the Landau-Zener effect [25], while $\hat{N}_1$ and $\hat{N}_2$ are particles numbers operators acting in the regions of spaces associated to the two nascent fragments. The condition obtained by means of the Lagrange multiplier λ produces a dynamical projection of the numbers of particles on the two final fragments [26]. The trial wave functions $|\varphi_{IM}\rangle$ are considered as a superposition of seniority-1 Bogoliubov wave functions. After some algebras, the equations of motion follow [24, 27]:

$$\begin{aligned}
-i\hbar\dot{c}^*_{\Omega,m} = c^*_{\Omega,m} \Big\{ & 2\textstyle\sum_{(\Omega',m')\neq(\Omega,m)} \mid v_{\Omega',m'(\Omega,m)} \mid^2 (\epsilon_{\Omega',m'} - \lambda) + (\epsilon_{\Omega,m} - s_k\lambda N_k) \\
& -G \left| \textstyle\sum_{(\Omega'm')\neq(\Omega,m)} u_{\Omega',m'(\Omega,m)} v_{\Omega',m'(\Omega,m)} \right|^2 - G \textstyle\sum_{(\Omega',m')\neq(\Omega,m)} \mid v_{\Omega',m'(\Omega,m)} \mid^4 \Big\} \\
& + \tfrac{\hbar^2}{2J} c^*_{\Omega,m} \left[I(I+1) - \Omega^2 \right] - \tfrac{\hbar^2}{2J} \Big\{ \textstyle\sum_{m'} c^*_{\Omega+1,m'} ((I-\Omega)(I+\Omega+1))^{1/2} \\
\times \Big[u_{\Omega,m(\Omega+1,m')} u_{\Omega+1,m'(\Omega,m)} & + v^*_{\Omega,m(\Omega+1,m')} v_{\Omega+1,m'(\Omega,m)} \Big] \langle \Omega+1, m' | j_+ | \Omega, m \rangle \, T_{\Omega+1,m',\Omega,m} \\
& + \textstyle\sum_{m'} c^*_{\Omega-1,m'} ((I+\Omega)(I-\Omega+1))^{1/2} \\
\times \Big[u_{\Omega,m(\Omega-1,m')} u_{\Omega-1,m'(\Omega,m)} & + v^*_{\Omega,m(\Omega-1,m')} v_{\Omega-1,m'(\Omega,m)} \Big] \langle \Omega-1, m' | j_- | \Omega, m \rangle \, T_{\Omega-1,m',\Omega,m} \Big\} \\
& -i\hbar c^*_{\Omega,m} \textstyle\sum_{(\Omega',m')\neq(\Omega,m)} \tfrac{1}{2} \left(v^*_{\Omega',m'(\Omega,m)} \dot{v}_{\Omega',m'(\Omega,m)} - \dot{v}^*_{\Omega'm'(\Omega,m)} v_{\Omega',m'(\Omega,m)} \right) \\
& + \textstyle\sum_{m'\neq m} h_{\Omega,m',m} c^*_{\Omega,m'} T_{\Omega,m',\Omega,m}
\end{aligned} \tag{2}$$

for the configuration mixing and

$$\begin{aligned}
-i\hbar\dot{v}^*_{\Omega',m'(\Omega,m)} = {} & 2v^*_{\Omega',m'(\Omega,m)} (\epsilon_{\Omega',m'} - \lambda) - G \textstyle\sum_{(\Omega'',m'')\neq(\Omega,m)} \Big\{ u_{\Omega'',m''(\Omega,m)} v^*_{\Omega'',m''(\Omega,m)} \\
& \times \left(u_{\Omega',m'(\Omega,m)} - \frac{v_{\Omega',m'(\Omega,m)} v^*_{\Omega',m'(\Omega,m)}}{2u_{\Omega',m'(\Omega,m)}} \right) - u_{\Omega'',m''(\Omega,m)} v_{\Omega'',m''(\Omega,m)} \frac{v^*_{\Omega',m'(\Omega,m)} v^*_{\Omega',m'(\Omega,m)}}{2u_{\Omega',m'(\Omega,m)}} \Big\} \\
& -2G v_{\Omega',m'(\Omega,m)} v^*_{\Omega',m'(\Omega,m)} v^*_{\Omega',m'(\Omega,m)}.
\end{aligned} \tag{3}$$

for the time-dependent pairing equations. In Eq. (2), I is the total spin of the nuclear system, Ω is the projection of the intrinsic spin on the axial axis of symmetry, m is a number that indexes the single-particle level in the pairing active levels space, λ is a Lagrange multiplier, s_k has values ± 1 contributing to the dynamical projection of the number of particles N_k of the nascent fragments, G is the pairing strength, J is the moment of inertia calculated with the cranking approximation, T_γ are overlaps between Bogoliubov wave functions. The mixing between two seniority-1 configurations having the unpaired nucleons located in the states $(\Omega \pm 1, m)$ and (Ω, m') that differ by one unit in the projection of the intrinsic spin is produced by means of the matrix elements of the Coriolis coupling $\langle \Omega \mp 1, m' | j_\mp | \Omega, m \rangle$. The mixing between two single particle levels having the same good quantum numbers Ω is realized by means of the Landau-Zener interaction strength $h_{\Omega,m,m'}$. The parameter $c_{\Omega,m}$ is the amplitude of the seniority-1 wave function in which the single particle level (Ω, m) is blocked. The parameters $u_{\Omega,m(\Omega',m')}$ and $v_{\Omega,m(\Omega',m)}$ are the BCS vacancy and occupation amplitudes of the single particle level (Ω, m) in the seniority-1 configuration (Ω', m'). The Rels. (3) are the well known time-dependent pairing equation, formally similar to the time-dependent Hartree-Fock equations [28, 29]. By including the Landau-Zener promotion mechanism, these equations were able to explain microscopically the energy partition in low energy fission [26, 30] or the odd-even effect in cold fission [31, 32].

3 Results

The fine structure of ^{211}Po and ^{211}Bi α-decays are reported, the daughter nuclei being ^{207}Pb and ^{207}Ta, respectively. A superasymmetric fission path between the ground state of the parent nucleus and the scission configuration is obtained according to the least action principle. In this context, the penetrability in the WKB approximation

$$P = \exp\left\{-\frac{2}{\hbar}\int_{R_1}^{R_2} \sqrt{2V(R)B(R)}dr\right\} \tag{4}$$

is minimized. Two ingredients are needed to perform the minimization, the deformation energy $V(R)$ and the effective mass $B(R)$. The deformation energy is obtained in the framework of the macroscopic-microscopic model [33] extended for binary system with different charge densities while the inertia is computed within the cranking approach [34–36]. To avoid the singularities in the inertia during the passage of the avoided crossing regions, the matrix elements of the Hamiltonian between the state of the blocked level and the paired level states are neglected. Recent formulas for odd-mass inertia can be found in Refs. [36, 37], which differ slightly between them but both produce singularities. The surface term of liquid drop part of the deformation energy is obtained with the Yukawa plus exponential model. The shell and pairing effects are obtained with the Strutinsky prescriptions [34] based on the Woods-Saxon two-center shell model [25]. A smooth transition of the whole system between the one-body configuration and a two-body one is managed by our nuclear shape parametrization. Our nuclear shape parametrization is determined by five generalized coordinates: the distance between the centers of the nascent fragments denoted R in the following, the curvature of the neck, the mass-asymmetry and the deformations of both fragments. A transition between the shell and pairing effects of the parent nucleus to those of both individual nuclei is realized in the scission region where the distance between the centers of the fragments is about $R \approx 9$ fm, as already made for the α-decay of ^{296}Lv in Ref. [38].

The barriers for the α-decay of the two investigated parent nuclei are plotted in Fig. 1. A molecular minimum is produced for a distance between the centers of the fragments $R \approx 7$ fm, which is due to the appearance of a strong shell effects produced by the reorganization of the single particle levels in a way compatible to the formation of the daughters, as explained previously in Refs. [20, 38]. The daughters fail to reproduce the magic numbers by only one nucleon missing. A such molecular minimum was also noticed in Hartree-Fock treatments of α-decay [16, 39]. The barriers of α-decay resemble with the double humped fission barriers

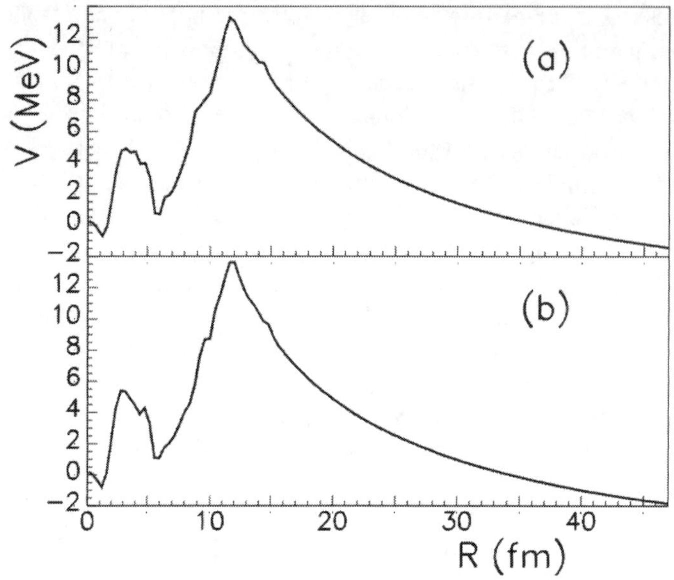

Figure 1. The α-decay superasimmetric fission barriers are plotted with for ^{211}Po in panel (a) and for ^{211}Bi in panel (b). The barriers are renormalized for an zero point kinetic energy of 0.5 MeV in the ground state. Figure compiled from Refs. [24, 27].

Table 1. Comparison between experimental and theoretical transition probabilities P_x on different final configurations x = 0,1,2 of the daughter nucleus. The index 0,1,2 mean the ground state, the first excited state and the second excited state, respectively. The orbitals of the daughter nucleus blocked by the unpaired nucleon for different values of x are given in a separated row. The first column gives the parent nucleus involved in the α-decay process. The following three columns addresses experimental transition probabilities (Exp) together with the final orbital for the lowest energy states. The last three columns are reserved for theoretical results (The). The experimental data are from Refs. [49, 50].

Parent nucleus	P_0 (Exp)	P_1 (Exp)	P_2 (Exp)	P_0 (The)	P_1 (The)	P_2 (The)
^{211}Po	$3p_{1/2}$	$2f_{5/2}$	$3p_{3/2}$	$3p_{1/2}$	$2f_{5/2}$	$3p_{3/2}$
^{211}Po	98.9%	0.55%	0.54%	99.1%	0.26%	0.62%
^{211}Bi	$3s_{1/2}$	$2d_{3/2}$	$1h_{11/2}$	$3s_{1/2}$	$2d_{3/2}$	$1h_{11/2}$
^{211}Bi	83.77%	16.23%	<0.0018%	87.9%	12.1%	6×10^{-15}

as postulated by theories [40–43]. Therefore, it can be predicted that β-vibrational states [44, 45] are produced in the second potential well, explaining some anormal transition states observed experimentally [46, 47]

The microscopic equations of motion (2) and (3) are resolved along the superasymmetric fission path for the α-decay of the two parent nuclei as detailed in Refs. [24, 27]. The Coriolis couplings and the single particle energies were determined within the two center Woods-Saxon wave functions. The initial states are $[\pi(h_{9/2})^2\nu(g_{9/2})^1]_{9/2}$ and $[\pi(h_{9/2})^1\nu(g_{9/2})^2]_{9/2}$ for ^{211}Po and ^{211}Bi, respectively, the total spin I being 9/2. The momentum of inertia was computed in the framework of the cranking model. The Landau-Zener avoided crossings regions were identified in a consistent way. In a true avoided levels crossing region, as discussed in Ref. [48], the levels should exchange their characteristics. So, in each region of closest approach between two levels with the same good quantum numbers, an avoided levels crossing region was confirmed by the exchange of the intrinsic angular momenta at the same deformation. The magnitudes of the Landau-Zener interactions were extracted from the distances of closest approach between two adiabatic levels. Therefore, all the ingredients needed to solve the equations of motion are available. Only the internuclear velocity $\dot{R}$ is missing. The internuclear velocity is considered as a fitting parameter. For a initial configuration, the equations of motion provide all the possible final configurations and their probabilities of realization at scission. For each configuration characterized by an orbital angular momentum and a single particle excitation, I calculated the barrier penetrability. The products of configuration probabilities and penetrabilities are combined for each final orbital of the quasi-spherical daughter giving probabilities of transitions on different states, after a proper normalization. The best agreement between theory and experiment was obtained for an internuclear distance velocity of $\dot{R} = 2\times10^4$ fm/fs. A comparison between experimental data and theoretical results is given in table 1. At this velocity, the dissipated energy can be neglected [51]. Other microscopic models predict larger velocities [52].

4 Conclusions

The fine structure of α-decay can be investigated microscopically with fission-like theories. A superasymmetric fission path compatible with the α-decay process was obtained. The microscopic equations of motion, extended to include configuration mixing mechanisms, were resolved. The probabilities to obtain different configurations at scission were calculated by starting from the ground state of the parent nucleus. This probabilities offer the possibility to evaluate the fine structure characteristics for α-decay. The theoretical results succeeded to reproduce well the experimental data.

This work was supported by the grant of Ministery of Research and Innovation, CNCS-UEFISCDI, project numbers PN-III-P4-ID-PCE-2016-0092, within PNCDI III.

References

[1] H.J. Mang, Z. Phys. **148**, 582 (1957)
[2] H.J. Mang, Ann. Rev. Nucl. Sci. **14**, 1 (1964)
[3] A. Sandulescu and M. Iosifescu, Nucl. Phys. **26**, 209 (1961)
[4] D. Deng, Z. Ren, D. Ni, and Y Qian, J. Phys. G **42**, 075106 (2015)
[5] D.S. Delion, Z.Z. Ren, A. Dumitrescu, and D.D. Ni, J. Phys. G **45**, 053001 (2018)
[6] W.M. Seif and A. Adel, Phys. Rev. C **99**, 044311 (2019)
[7] A. Adel and T. Alharbi, Nucl. Phys. A **975**, 1 (2018)
[8] V. Dehghami, S.A. Alavi, and Kh. Benam, Mod. Phys. Lett. A **33**, 1850080 (2018)
[9] G.M. Carmel Viola Bai and R. Nithya Agnes, Pramana **93**, 39 (2019)
[10] A.O. Silisteanu, C.I. Anghel, and I. Silisteanu, Rom. J. Phys. **63**, 302 (2018)
[11] H.M. Devajara, Y.K. Gambir, A. Bhagwat, M. Gupta, S. Heinz and G. Munzenberg, Rom. J. Phys. **63**, 304 (2018)
[12] A. Soylu, Chin. Phys. C **43**, 074102 (2019)
[13] M. Ismail and A. Adel, J. Phys. G **44**, 125106 (2017)
[14] D.S. Delion, Phys. Rev. C **80**, 024310 (2009)
[15] K.P. Santhosh, J.G. Joseph and B. Priyanka, Nucl. Phys. A **877**, 1 (2012)
[16] C. Xu, G. Röpke, P. Schuck, Z. Ren, Y. Funaki, H. Horiuchi, A. Tohsaki, T. Yamada, and B. Zhou, Phys. Rev. C **95**, 061306(R) (2017)

[17] G. Röpke, P. Schuck, Y. Funaki, H. Horiuchi, Z. Ren, A. Tohsaki, C. Xu, T.Yamada, and B. Zhou, Phys. Rev. C **90**, 034304 (2014)
[18] G. Röpke, P. Schuck, C. Xu, Z. Ren, M. Lyu, B. Zhou, Y. Funaki, H. Horiuchi, A. Tohsaki, and T. Yamada, J. Low Temp. Phys. **189**, 383 (2017)
[19] D.S. Delion, A. Dumitrescu, and V.V. Baran, Phys. Rev. C **97**, 064303 (2018)
[20] M. Mirea, Eur. Phys. J. A **51**, 36 (2015)
[21] M. Mirea, R. Budaca, and A. Sandulescu, Ann. Phys. **380**, 154 (2017)
[22] A. Sandulescu and M. Mirea, Eur. Phys. J. A **50**, 110 (2014)
[23] D.L. Hill and J.A. Wheeler, Phys. Rev. **89**, 1102 (1953)
[24] M. Mirea, Phys. Rev. C **96**, 064607 (2017)
[25] M. Mirea, Phys. Rev. C **78**, 044618 (2008)
[26] M. Mirea, Phys. Rev. C **89**, 034623 (2014)
[27] M. Mirea, EPL **124**, 12001 (2018)
[28] S.E. Koonin and J.R. Nix, Phys. Rev. C **13**, 209 (1976)
[29] J. Blocki and H. Flocard, Nucl. Phys. A **273**, 45 (1976)
[30] M. Mirea, Phys. Lett. B **717**, 252 (2012)
[31] M. Mirea, Phys. Lett. B **680**, 316 (2009)
[32] M. Mirea, Phys. Rev. C **83**, 054608 (2011)
[33] P. Moller, J.R. Nix, W.D. Myers, and W.J. Swiatecki, At. Data. Nucl. Data. Tables **59** 185 (1995)
[34] M. Brack, J. Damgaard, A. Jensen, H. Pauli, V. Strutinsky, and W. Wong, Rev. Mod. Phys. **44**, 320 (1972)
[35] M. Mirea, J. Phys. G **43**, 105103 (2016)
[36] M. Mirea, Phys. Rev. C **100**, 014607 (2019)
[37] W. Brodzinski, M. Kowal, J. Skalski, and P. Jachimowicz, Acta Phys. Pol. B **49**, 621 (2018)
[38] A. Sandulescu, M. Mirea, and D.S. Delion, EPL **101**, 62001 (2013)
[39] G.G. Adamian, N.V. Antonenko, H. Lenske, S.V. Tolokonnikov, and E.E. Saperstein, Phys. Rev. C **94**, 054309 (2016)
[40] K.-H. Schmidt and B. Jurado, Rep. Prog. Phys. **81**, 106301 (2018)
[41] M. Mirea, Int. J. Mod. Phys. E **27**, 1850076 (2018)
[42] Zhiming Wang, Zhong Zhu, CL. Zhong, and TS. Fan, Nucl. Phys. A **989**, 81 (2019)
[43] Bing-Nan Lu, E.G. Zhao, and S.G. Zhou, Phys. Rev. C **85**, 011301 (2012)
[44] S. Bjornholm and J.E. Lynn, Rev. Mod. Phys. **52**, 725 (1980)
[45] M. Mirea, Rom. J. Phys. **64**, 305 (2019)
[46] A. Astier, P. Petkov, M.-G. Porquet, D.S. Delion, and P. Schuck, Eur. Phys. J. A **46**, 165 (2010)
[47] D.S. Delion and R.J. Liotta, Phys. Rev. C **87**, 041302(R) (2013)
[48] C. Zener, Proc. R. Soc. A **137**, 696 (1932)
[49] L.J. Jardine, Phys. Rev. C **11**, 1385 (1975)
[50] F.G. Kondev and S. Lalkovski, Nucl. Data Sheets **112**, 707 (2011)
[51] M. Mirea and A. Sandulescu, Rom. Rep. Phys. **70**, 201 (2018)
[52] O. Serot, N. Carjan, and D. Strottman, Nucl. Phys. A **569**, 562 (1994)

Investigation of fission product isomeric ratios and angular momenta of ^{132}Sn populated in the ^{241}Pu(n_{th},f) reaction

Jehaan Nicholson[1], *Abdelhazize* Chebboubi[1,*], *Olivier* Serot[1], *Grégoire* Kessedjian[2], *Yung Hee* Kim[3], *Ulli* Köster[3], *Olivier* Litaize[1], *Olivier* Méplan[2], and *Christophe* Sage[2], and *Mourad* Ramdhane[2]

[1]CEA, DEN, DER, SPRC, Cadarache, Physics Studies Laboratory, 13108 Saint-Paul-lès-Durance, France

[2]LPSC, Université Grenoble-Alpes, CNRS/IN2P3, 38026 Grenoble Cedex, France

[3]Institut Laue-Langevin, 38042 Grenoble Cedex 9, France

Abstract. During an experimental campaign performed at the LOHENGRIN recoil spectrometer of the Institut Laue-Langevin (ILL), a kinetic energy dependence of ^{132}Sn fission product isomeric ratio (IR) has been measured by inducing thermal fission of ^{241}Pu. The IRs are deduced using gamma ray spectrometry in coincidence with the ionisation chamber. To interpret these data, we use the FIFRELIN Monte-Carlo code to simulate the de-excitation of the fission fragments. Combining the IRs with the FIFRELIN calculations, the angular momentum distribution with kinetic energy of the doubly magic nucleus of ^{132}Sn was deduced. This will be compared with the angular momentum distribution obtained for the reaction ^{235}U(n_{th},f) for ^{132}Sn.

1 Introduction

The past decade has seen a growing energy need and thus a renewed interest in nuclear energy. Nonetheless, to make the new reactors more safe, the existing technology needs to be improved and the challenge for new and innovative fuel must be overcome. Furthermore, a precise understanding of the fission process is a noteworthy challenge faced by nuclear physicists even though eight decades have passed since its discovery in 1939 [1, 2]. Many different models and hypotheses such as the liquid drop model, shell model etc. have been developed to explain and reproduce the experimental data obtained as well as to improve the understanding of fission. However, the angular momenta of the fission fragments are a poorly known quantity. Angular momenta of fission fragments are a component of the phenomenological model used to assess nuclear observables used in applications as prompt gammas and neutrons. This angular momentum can be estimated by using models, see for example Ref. [3], or through direct measurements, see Ref. [4] and references therein. On the other hand, one can study indirectly the angular momenta of the fission fragments by measuring the isomeric ratios of the fission fragments [5–7].

*e-mail: Abdelhazize.CHEBBOUBI@cea.fr

2 Experiment

The experiment was conducted at the LOHENGRIN recoil spectrometer [8] at the Institut Laue-Langevin (ILL) [9]. The LOHENGRIN recoil spectrometer was built in the 1970s and is one of the key instruments at the ILL to carry out experiments for nuclear physics and nuclear data. The LOHENGRIN recoil spectrometer enables us to study the mass, charge and kinetic energy distribution of the fission products. These fission products are obtained by exposing a fissile or fertile target to a thermal neutron flux of about 5×10^{14} n/cm^2/s near the core of the ILL high-flux reactor. The total length of LOHENGRIN is 23 m and has a high mass resolution ($A/\Delta A \cong 400$) [10], which is dependent on target size. By using the electric and magnetic fields of LOHENGRIN, fission products of interest are extracted and brought to the detection area. These fields are perpendicular to each other and have focussing properties in their respective planes.

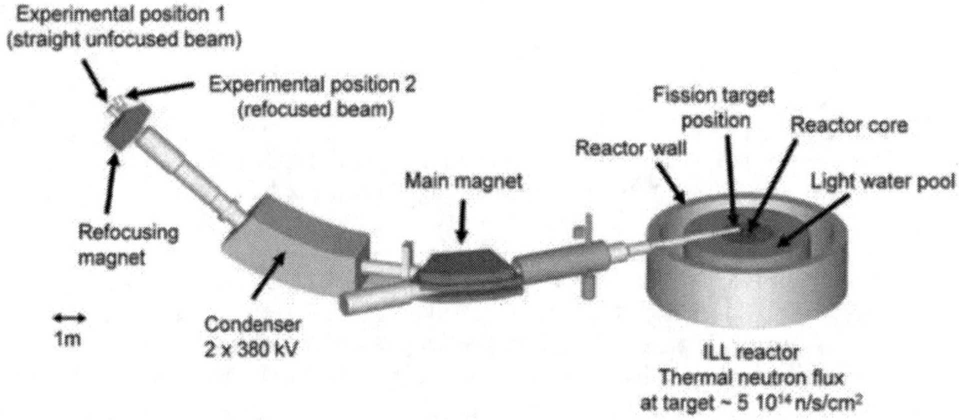

Figure 1. Schematic of the LOHENGRIN mass spectrometer showing the dipole magnet in blue (centre) and the electrostatic condenser in yellow. They are used to generate the electric and magnetic fields, which are used to select and divert the charged fission products.

For this experiment, we used ^{241}Pu (39.9 μg/cm^2) as the target nuclei. By varying the electric and magnetic fields, the desired fission products are selected by the A/q and E_k/q ratio, where A is the desired mass of the fission product, q is the ionic charge and E_k its kinetic energy. As can be seen from Fig. 1, there are two experimental positions present. At experiment position 1, the beam has an energy dispersion of 7.2 cm per 1% energy difference, whereas at position 2, the beam is refocussed in the energy axis using the RED magnet [11] and, hence, the particle flux density is increased by up to a factor of seven [10] as compared to position 1. In our experiment, the detectors were set up at position 2. The detection setup consists of an ionisation chamber and two HPGe clover detectors with four germanium crystals each. This is depicted in Figure 2.

The extracted fission products are then implanted in the Al foil at the top of the ionisation chamber. The gamma rays, emitted by internal transition (IT) of the isomeric state and after the β- decay of the ground state are detected using the two clover detectors. By using LOHENGRIN, ^{132}Sn was extracted at ionic charges 20 and 24 respectively with kinetic energy ranging from 57 MeV to 84 MeV.

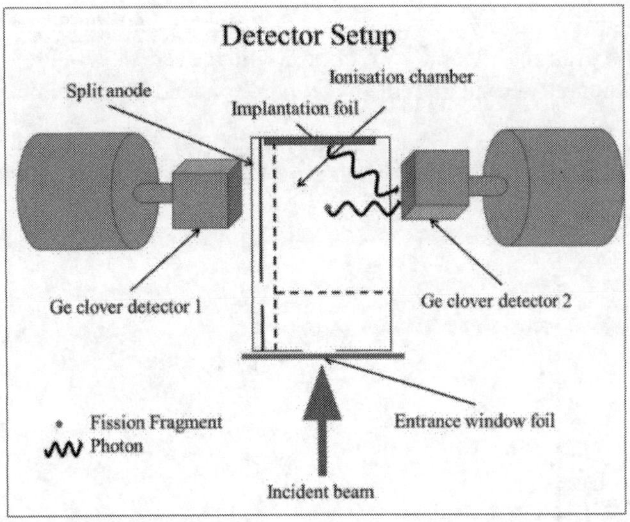

Figure 2. Detector setup showing the two HPGe clover detectors in green and the split-anode ionisation chamber in between them. The implantation foil is placed at the top of the ionisation chamber to trap the fission products.

3 Isomeric ratios from experiments

Isomeric ratios (IR) are defined as ratio of the production rate (P_i) of one isomeric state to the sum of the production rates of all the isomeric states and the ground state (GS).

$$IR = \frac{P_i}{\sum P_i + GS} \tag{1}$$

3.1 The analysis

^{132}Sn has one microsecond isomeric state at 4848.5 keV having $J^{\pi} = 8^{+}$ with a half-life of 2.08 μs [12]. The analysis was carried out off-line. To analyse the gamma rays originating from the isomeric state, a coincidence spectrum was generated. The time coincidence window was set to 20 μs. The coincidence was created between the ionisation chamber and the HPGe clover detectors. By using the coincidence method to generate a spectrum, we are able to reduce the gammas originating from room background and beta decays and gammas from β- decay of the ground state of the parent nucleus (^{132}In). Thus, we obtain a clean spectrum for the gamma rays originating from the microsecond isomeric state of ^{132}Sn as seen in the Fig. 3. For the measurement of the gamma rays originating from the β- decay of the ground state (^{132}Sn), the ungated spectrum was used.

The gamma spectrum was analysed using the TV gamma spectrum analyser program [13]. The intensities of the gamma rays were obtained from literature [12] and the efficiencies for the germanium detectors were obtained from simulations validated with experimental data. To calculate the IR from the counts obtained from the gamma spectra, firstly, we have to calculate the production rates of the isomeric state as well as the ground state. To do this, the Bateman equations need to be solved. Along with this, corrections for the decay during flight also need to be taken into account; this is due to the fact that the distance between target and the experimental focal plane is 23 m. To accurately determine the production rates and the IR along with their respective uncertainties, a Monte-Carlo method [5, 14] is being used. This

Monte-Carlo Code (MCC) is used as compared to the analytical method as there are quite a few parameters involved in different corrections, which need to be added. Hence trying to propagate and calculate their uncertainties analytically would be a complex task.

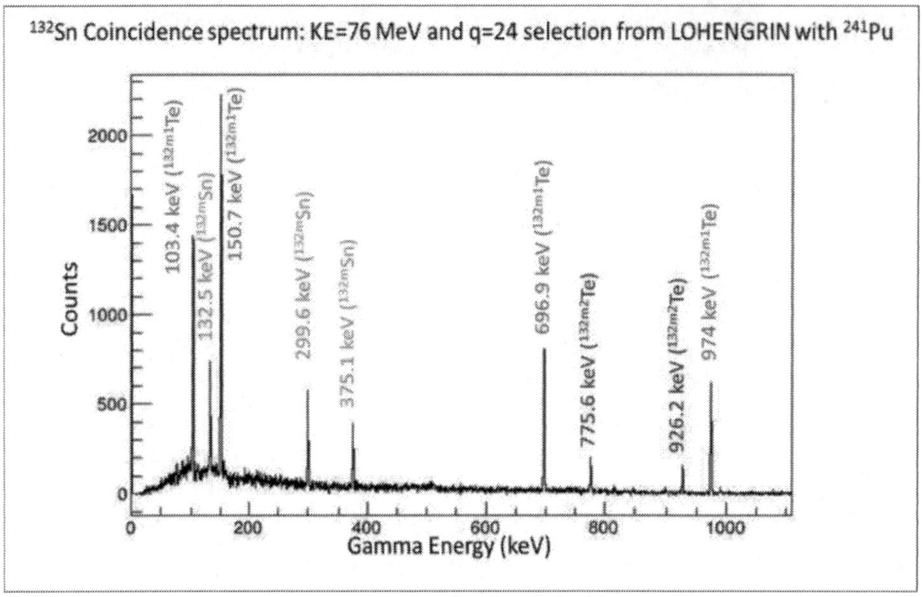

Figure 3. Gated Spectrum showing the three gamma rays of interest (in green) from the ^{132}Sn isomeric state (IS). Gamma rays from the two ISs of ^{132}Te are also observed in this spectrum.

As input to the MCC, one must provide the half-life of the states (ground and isomeric), counts and the error on counts which are extracted using TV gamma spectrum analyser, detection efficiency, the gamma intensities and the normalisation factor for the gamma intensities, the branching ratio, kinetic energy selected by LOHENGRIN and the flight path length. The MCC then calculates the average number of disintegrations (N_d), the production rates and the isomeric ratios along with their respective uncertainties. One can also obtain sensibility plots as well as covariance matrices from the MCC.

In the MCC, N_d is calculated for each of the gamma rays (N_{di}) arising from a particular state, these N_{di} are then used to calculate the mean N_d. The χ^2 test is used to verify whether a 90% level of confidence is achieved. If not, progressively, uncertainty is added [15] and the N_d is recalculated to achieve the test criteria. Thus, a final mean value of the N_d along with its uncertainty is obtained. The systematic uncertainties account for less than 5%, whereas the statistical uncertainties go as high as 50%.

The beamtime had a span of ten days and hence, not all the measurements were performed consecutively or even on the same day. For this reason, we must take into account the target evolution. To see the target evolution, several energy scans were carried out throughout the experiment schedule. The shift in the mean value of the kinetic energy between the first and the last experimental days is equal to (4.6 ± 1.0) MeV. It should be noted that the evolution of the mean kinetic energy was linear with a slope of (−0.52 ± 0.09) MeV/day. For each kinetic energy selected with the LOHENGRIN spectrometer, IRs are measured. By combining all the experimental data (weighted by the kinetic energy distribution) the mean IR can be derived and is equal to 0.0719 ± 0.0016. It should be noted that covariance was not taken into account for the uncertainty propagation.

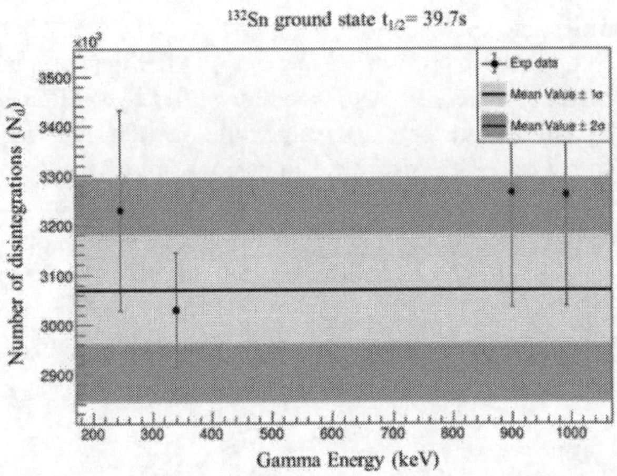

Figure 4. Number of disintegrations calculation using the MCC for GS for 72 MeV and q=24 selection of LOHENGRIN.

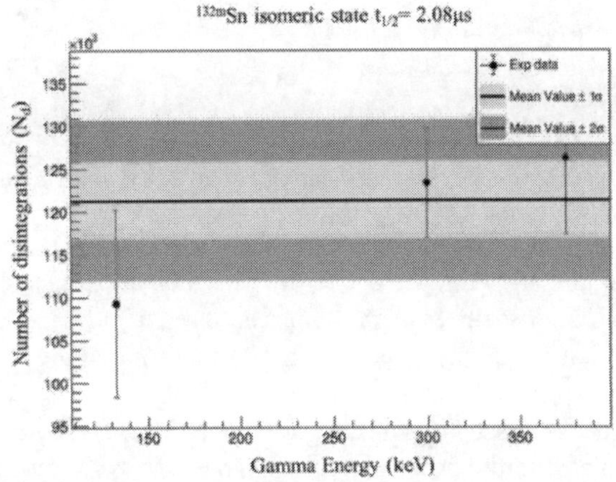

Figure 5. Number of disintegrations calculation using the MCC for IS for 72 MeV and q=24 selection of LOHENGRIN.

4 Spin extraction using FIFRELIN

FIssion FRagment Evaporation Leading to an Investigation in Nuclear data (FIFRELIN) [16] is a Monte-Carlo code developed at CEA. This code simulates the fission process and gives information on fission observables such as prompt neutrons, gammas, neutron multiplicity. For this work, this code has only been used to carry out the de-excitation simulation of the nuclei of interest (^{132}Sn). The inputs provided for this code are the mass number (A), the atomic number (Z), excitation energy (E^*) and spin along with parity (J^π). It also requires additional files from the RIPL-3 2015 [17, 18] library to get information on the nuclear levels and the gammas and electrons emitted at lower energies. Additional models are also required such as the Composite Gilbert-Cameron Model (CGCM), which is propositioned in RIPL-3 [17] for nuclear level densities, Back Shifted Fermi Gas Model (BSFGM) for spin cut-off and the Enhanced General Lorentzian (EGLO) model [19] for the gamma strength functions. The BRICC code [20] is used to calculate the internal conversion coefficients. FIFRELIN uses these input libraries and assumes that the experimental level scheme provided in the RIPL-3 2015 file is complete up to cut-off energy (E_{cut}). After this E_{cut}, it is assumed that the experimental level scheme is incomplete and this is then filled by FIFRELIN using the CGCM model. The BSFGM is used to attribute a J^{π} value to these new levels and the EGLO model is used to get the gammas and their respective intensities originating from these levels. The probability to emit neutrons are obtained from the neutron transmission coefficients which are derived from an optical model; the Koning-Delaroche global neutron optical model potential [21] but in our case, no neutrons are emitted as the simulations were run below neutron separation energy.

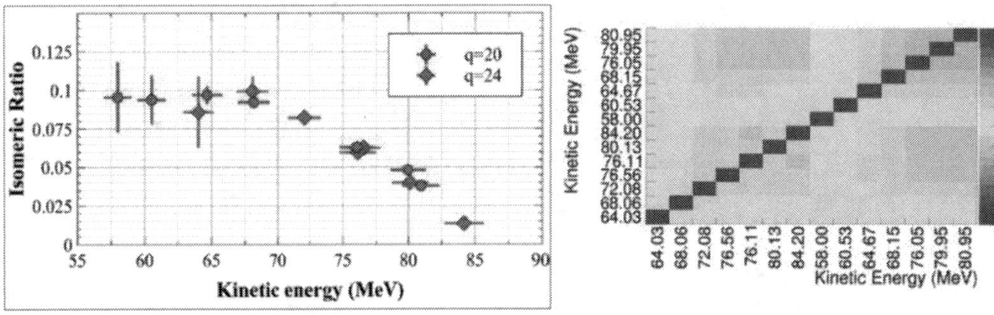

Figure 6. Isomeric Ratio of ^{132}Sn from thermal neutron induced fission of ^{241}Pu measured at two different ionic charge selections (left) and the associated covariance matrix (right). The kinetic energies are corrected from the relative evolution of the energy loss during the experimental campaign.

Once the input parameters, A, Z, E^* and J^{π} are provided, a cascade can start. The above-mentioned models and experimental level schemes are used to calculate the isomeric ratio for a particular E^* and J^{π} combination. The E^* has a range starting at the energy of the isomeric state up to the neutron separation energy. For each excitation energy, a range J^{π} of values ($0^{\pm}$ to $30^{\pm}$) is given. For each of these J^{π} values, an IR ($IR_{FIF}(E^*, J^{\pi})$) is calculated by FIFRELIN. To compare these results with the experimental data, the results are averaged by the equation given below:

$$IR_{FIF}(E^*, J_{rms}) = \sum_J \sum_\pi P(\pi)P(J)IR_{FIF}(E^*, J^\pi) \tag{2}$$

$$\text{where } P(J) \propto (2J+1)\exp\left(-\frac{\left(J+\frac{1}{2}\right)^2}{J_{rms}^2}\right) \tag{3}$$

$$\text{and } P(\pi) = P(\pm 1) = \frac{1}{2} \tag{4}$$

The Likelihood method is used to adjust the spin cut-off (J_{rms},) which in that case is a free parameter

$$L(E^*, J_{rms} \mid E_k) \propto \exp\left(-\frac{\left(IR_{exp}(E_k) - IR_{FIF}(E^*, J_{rms})\right)^2}{2\left(\sigma_{exp}^2, \sigma_{FIF}^2\right)}\right) \tag{5}$$

where, $IR_{exp}(E_k)$ and σ_{exp} are the isomeric ratios and their uncertainty obtained from the experiments that is dependent on the selected kinetic energy from LOHENGRIN. σ_{FIF} is the uncertainty obtained from FIFRELIN.

5 Results and conclusion

Combining the FIFRELIN calculations with the experimental results, we were able to extract the J_{rms} value as a function of kinetic energy. Figure 7 depicts the results from this work, which have been compared to the results obtained by using a ^{235}U target [5]. It can be seen that the J_{rms} value obtained from two different fissioning systems are quite similar. Also for this work, one can observe a flat plateau-like region at lower kinetic energies. Further experiments and calculations need to be carried out to explain this phenomenon. Experimental results from this work have been further compared with the calculations using the Madland-England (M.E.) model and the GEF code. This can be seen in Table 1. The M.E. model uses the assumption that the isomeric ratio is only dependant on the spin of both the ground state and the isomeric state. Furthermore, it works on the supposition that all the fission fragments are characterised by a spin cut-off value of (7.5±0.5) h, which gives an isomeric ratio of 0.642±0.039 for ^{132}Sn (M.E. (a)). By using the isomeric ratio from this work, which is 0.0719±0.0016, a J_{rms} of 2.8±0.1 (M.E. (b)) is obtained from the M.E. model. We clearly observe a mismatch between the experimental results and those obtained from the M.E. model for ^{132}Sn.

Table 1. Comparison of results: In M.E. (a), $\overline{\mathbf{J}_{rms}}$ is set to 7.5 h and the IR is calculated by using M.E. model. In M.E. (b), the IR is set equivalent to our experimental result and the $\overline{\mathbf{J}_{rms}}$ is calculated using the same model.

Experiments	$\overline{IR}$	$\overline{J_{rms}}$ ($\hbar$)
This work (^{241}Pu)	0.0719±0.0016	4.8±0.1
^{235}U [5]	0.054±0.006	4.7±0.2
Models		
Madland-England (a)	0.642±0.039	7.5±0.5
Madland-England (b)	0.0719±0.0016	2.8±0.1
GEF [22]	0.234	6.65±0.03

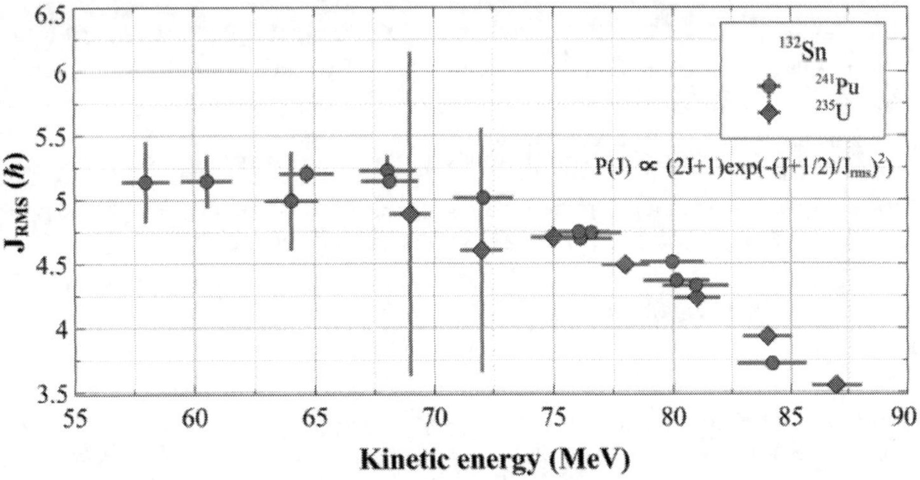

Figure 7. Spin of ^{132}Sn and its dependence on kinetic energy using ^{241}Pu and ^{235}U targets. The kinetic energies are corrected from the relative evolution of the energy loss during the experimental campaign.

In conclusion, the dependence of the isomeric ratios of the fission products on their kinetic energy was obtained. Using statistical analysis, along with FIFRELIN calculations involving level density models, gamma strength functions, spin cut-off models and internal conversion coefficients, we were able to determine the J_{rms} value for each of the isomeric ratios obtained. Furthermore, it can be observed that the J_{rms} values for two different fissioning systems are quite similar and follow a similar trend. For nuclei such as ^{132}Sn, the results from the experiments should be taken into account for the nuclear data evaluations instead of using codes or models to compute them. In this same experimental campaign, experiments were carried out on other isotopes of Sn as well. It would be very interesting to see how the J_{rms} value changes with change in mass for the same element.

The authors would like to thank and express our gratitude towards the support staff of the ILL as well as the staff involved from CEA-Cadarache and LPSC Grenoble. This work has been supported by the NEEDS project, by CNRS and by CEA. This work has been done in collaboration with CEA-Cadarache, LPSC Grenoble and ILL.
Raw data of this experiment are available via ref. [23].

References

[1] O. Hahn, F.W. Strassmann, Naturwissenschaften **27**, 11 (1939)
[2] L. Meitner, O.R. Frisch, Nature **142**, 239 (1939)
[3] D. Madland, T. England, Nuclear science and Engineering **64**, 859-865 (1977)
[4] D. Tarrío, L.S. Leong, L. Audouin, I. Duran, C. Paradela, et al., Nuclear Instruments and Methods in Physics Research A **743**, 79-85 (2014)
[5] A. Chebboubi, G. Kessedjian, O. Litaize, O. Serot, H. Faust, D. Bernard, A. Blanc, U. Köster, O. Méplan, P. Mutti, C. Sage, Physics Letters B **775**, 190-195 (2017)
[6] A. Al-Adili, V. Rakopoulos, A. Solders, European Physical Journal A **55**, 61 (2019)
[7] V. Rakopoulos, M. Lantz, S. Pomp, A. Solders, A. Al-Adili, L. Canete, T. Eronen, A. Jokinen, A. Kamkainen, A. Mattera, I.D. Moore, D.A. Nesterenko, M. Reponen, S. Rinta-Antila, A. de Roubin, M. Vilén, M. Österlund, H. Penttilä, Physical Review C **99**, 014617 (2019)
[8] P. Armbruster, M. Asghar, J.P. Bocquet, R. Decker, H. Ewald, J. Greif, E. Moll, B. Pfeiffer, H. Schrader, F. Schussler, G. Siegert, H. Wollnik, Nuclear instruments and Methods **139**, 213-222 (1976)
[9] "About ILL; What is The ILL" Institut Laue-Langevin, [Online]. Available: https://www.ill.eu/about-ill/what-is-the-ill/
[10] "PN1 Fission-product spectrometer; PN1 characteristics" [Online]. Available: https://www.ill.eu/users/instruments/instruments-list/pn1/characteristics/
[11] G. Fioni, H.R. Faust, M. Gross, M. Hesse, P. Armbruster, F. Gönnenwein, Nuclear Instruments and Methods in Physics Research A **332**, 175-180 (1993)
[12] Y. Khazov, A.A. Rodionov, S. Sakharov, B. Singh, Nuclear Data Sheets **104**, 497-790 (2005)
[13] A. Fitzler, *Program TV,* Cologne: Institute for Nuclear Physics
[14] A. Chebboubi, *Contribution à l'étude de la fission nucléaire : de LOHENGRIN à FIPPS*, PhD thesis from Université de Grenoble (2015)
[15] S. Julien-Laferrière, A. Chebboubi, G. Kessedjian, O. Serot, EPJ Nuclear Sci. Technol., **4**, 25 (2018)
[16] O. Litaize, O. Serot, L. Berge, The European Physical Journal A **51**, 177 (2015)

[17] R. Capote, M. Herman, P. Oblozinsky, P.G. Young, S. Goriely, T. Belgya, A.V. Ignatyuk, A.J. Koning, S. Hilaire, V.A. Plujko, M. Avrigeanu, O. Bersillon, M. B. Chadwick, T. Fukahori, Z. Ge, Y. Han, S. Kailas, J. Kopecky, V.M. Maslov, G. Reffo, M. Sin, E. Sh. Soukhovitskii, P. Talou, Nuclear Data Sheets **110**, 3107-3214 (2009)

[18] M. Verpelli and R. Capote, "Structure of RIPL discrete level library files," [Online]. Available: https://www-nds.iaea.org/RIPL-3/levels/levels-readme.html

[19] J. Kopecky, M. Uhl, Physical Review C **41** (1990)

[20] T. Kibédi, T.W. Burrows, M.B. Trzhaskovskaya, P.M. Davidson and C.W. Nestor Jr, Nuclear Instruments and Methods in Physics Research A **589**, 202-229 (2008)

[21] A.J. Koning, J.P. Delaroche, Nuclear Physics A **713**, 321-310 (2003)

[22] K.H. Schmidt, B. Jurado, C. Amouroux and C. Schmitt, Nuclear Data Sheets **131**, 107-221 (2016)

[23] G. Kessedjian, D. Bernard, A. Blanc, A. Chebboubi, H. Faust, Y.H. Kim, U. Köster, O. Litaize, O. Méplan, J.N. Nicholson, M. Ramdhane, C. Sage and O. Serot, ILL, doi:10.5291/ILL-DATA.3-01-640 (2018)

Systematic study of the de-excitation of neutron-rich nuclei produced in different fission reactions

Andreas Oberstedt[1,2,*], *Angélique* Gatera[3], *Alf* Göök[3], *Matthieu* Lebois[4], *Stephan* Oberstedt[3], *Liqiang* Qi[4], *Christelle* Schmitt[5], and *Jonathan N.* Wilson[4]

[1]Extreme Light Infrastructure - Nuclear Physics (ELI-NP) / Horia Hulubei National Institute for Physics and Nuclear Engineering (IFIN-HH), 077125 Bucharest-Magurele, Romania
[2]Department of Physics and Astronomy, Uppsala University, 75120 Uppsala, Sweden
[3]European Commission, DG Joint Research Centre, Directorate G - Nuclear Safety and Security, Unit G.2 Standards for Nuclear Safety, Security and Safeguards, 2440 Geel, Belgium
[4]Université Paris-Saclay, CNRS/IN2P3, IJCLab, 91406 Orsay, France
[5]Institut Pluridisciplinaire Hubert Curien, 23 rue du Loess, B.P. 28, 67037 Strasbourg Cedex 2, France

Abstract. In this paper we present results from two recent studies, both related to the emission of prompt fission γ rays. Firstly, we have analyzed data from the reaction ^{235}U(n, f) induced by fast neutrons of average energy $\overline{E}_n = 1.7$ MeV. The deduced spectral characteristics are an average multiplicity $\overline{M}_\gamma = 7.11 \pm 0.44$ γ rays per fission and an average total γ-ray energy release in fission $\overline{E}_{\gamma,tot} = 5.51 \pm 0.46$ MeV, corresponding to an average γ-ray energy $\overline{\epsilon}_\gamma = 0.77 \pm 0.08$ MeV. Secondly, we have addressed – and answered – the question how those characteristics in general depend on the width of the chosen prompt time window and the timing resolution, determined by the employed detectors and electronics. The conclusion is that once this is known, it is possible to compare results from different experiments in a more meaningful way.

1 Introduction

The energy released in nuclear fission is distributed in kinetic and excitation energy of the two fragments. The excitation energy manifests itself in fragment deformation and intrinsic excitation energy. The first step of fission fragment de-excitation takes place within a few nanoseconds after scission through the successive emission of prompt neutrons. Then, in a simplified picture, when the fragments' remaining excitation energy is lower than the neutron binding energy, γ rays are emitted. Thereafter, de-excitation continues through β-decay towards isotopes in the valley of stability and subsequent (delayed) neutron and γ-ray emission [1, 2]. The prompt neutrons are important for maintaining a chain reaction in a reactor, while delayed neutrons allow reactor control. Prompt fission γ-ray emission contributes to the prompt heat released in fission, whereas the delayed γ-ray emission is counted as decay heat. Hence, the properties of the emitted particles prior to the onset of weak decays are important for both nuclear applications as well as for the better understanding of the mechanism of fission fragment de-excitation. While prompt fission neutron emission is being studied

*e-mail: andreas.oberstedt@eli-np.ro

during the last 50 years, prompt γ-ray emission is one of the least understood parts of the fission process and was investigated essentially in the early 1970s and then again since about a decade ago, and only for a limited number of fissioning isotopes (cf. Ref. [3] and references therein).

Concerning the latter, we have contributed in the past years with a number of precise measurements of prompt fission γ-ray spectra (PFGS) from various systems, i.e. the spontaneous fission of ^{252}Cf [4, 5] and ^{240,242}Pu [6], thermal-neutron induced fission on ^{235}U [7] and ^{239,241}Pu [8, 9], as well as fast-neutron induced fission on ^{235,238}U [10–12], and quite recently, ^{239}Pu [13]. For those we determined average characteristics like multiplicity, mean energy per photon and total γ-ray energy released in fission. The obtained results were investigated for their dependences of mass and atomic numbers of the fissioning system as well as the dissipated excitation energy. The purpose of this endeavor was to find a description that allows predicting prompt γ-ray spectra characteristics for cases that cannot be studied experimentally [14]. In this conference paper, however, we will present new results for PFGS characteristics from ^{235}U(n, f) and address the question how those characteristics in general depend on the width of the chosen time window and the timing resolution of the employed detectors and electronics. Finally, we will give an outlook on our future activities in this field.

2 Picking up the threads

In the very first experiment performed at the directional fast-neutron source LICORNE of IPN Orsay [15] some years ago, prompt-fission γ rays were measured with 14 hexagonal BaF_2 and three coaxial $LaBr_3$:Ce scintillation detectors. The γ rays were produced in fission reactions induced by neutrons of average energy $\overline{E}_n$ = 1.7 MeV hitting ^{235}U and ^{238}U samples of 10 mg approximate mass each, placed back-to-back at the central cathode position inside a cylindrical twin Frisch-grid ionization chamber. Until quite recently, only γ rays detected with the BaF_2 scintillators had been analyzed and first results were already published [10]. As stated there, the low counting statistics did not allow to treat the $LaBr_3$:Ce spectra back then due to difficulties in unfolding the detector responses. Concerning the BaF_2 spectra, the thresholds were rather high, approximately 500 keV. Moreover, during this experiment we encountered problems associated with dead time, which might have resulted in uncertainties related to a proper normalization of the γ-ray spectra with respect to the total number of occurred fission events. For the reasons mentioned above, no absolute values for PFGS characteristics (i.e. average multiplicity, average and total γ-ray energy) could have been deduced. Instead, only ratios for ^{235}U and ^{238}U spectral properties were given [10]. However, subsequent experiments have provided us now with the means for "picking up the thread" from this previous measurement and analyzing also the spectra taken with the $LaBr_3$:C detectors.

One of the prerequisites mentioned above is the recent measurement of prompt γ rays from the reaction ^{238}U(n, f) at $\overline{E}_n$ = 1.9 MeV [11]. Taking into account that the average prompt-fission γ-ray multiplicities for this reaction agree well within 1% for incident neutrons of 1.7 and 1.9 MeV kinetic energy, which may be inferred from the predictions in Ref. [14], the spectrum at 1.9 MeV can be used to obtain the correct normalization for the spectrum at 1.7 MeV by adjusting the latter to the first one. This was done for the region 0.5 MeV < E_γ < 6 MeV to reproduce the same multiplicity, and the result is shown in Fig. 1. Since both BaF_2 spectra for n + ^{235}U and n + ^{238}U, respectively, were detected simultaneously, the scaling factor obtained above for n + ^{238}U may be applied to the PFGS for n + ^{235}U. This in turn allows also to fix the absolute height of the ^{235}U(n, f) PFGS taken with the $LaBr_3$:Ce detectors to the one observed with the BaF_2 detectors.

As mentioned above, counting statistics was rather low for the $LaBr_3$:Ce detectors, in contrast to the spectra obtained with the massive BaF_2 clusters. For that reason a reliable

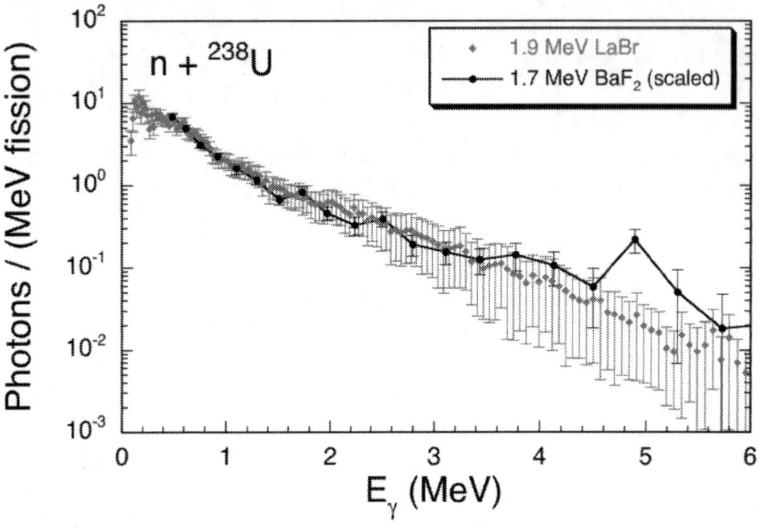

Figure 1. Comparison of the unfolded prompt-fission γ-ray spectra for n + ^{238}U at $\overline{E}_n$ = 1.9 MeV [11] and 1.7 MeV [10], where the latter was adjusted to the first one (see text for details).

unfolding of the detector response was not possible for the recorded $LaBr_3$:Ce spectra. However, we observed that the spectra measured in coincidence with the fission of n + ^{235}U and n + ^{238}U exhibit a very similar shape (details will be given elsewhere). Hence, instead of unfolding the measured ^{235}U(n, f) PFGS, we apply a transformation function, a procedure that was explained and applied previously to ^{240,242}Pu(sf) PFGS in Ref. [6]. In this case, the transformation function is obtained for the system n + ^{238}U, by using the measured $LaBr_3$:Ce spectrum at $\overline{E}_n$ = 1.7 MeV and the previously determined emission spectrum at $\overline{E}_n$ = 1.9 MeV. Here we apply the same argument as above that the emission spectra at 1.7 and 1.9 MeV are practically identical, which is corroborated by Fig. 1, at least for γ energies below 3 MeV. However, this does not imply any restriction, because the low statistics for the 1.7 MeV measurement does not allow to consider $LaBr_3$:Ce spectra above that energy anyway. As a result, we obtain emission spectra for n + ^{235}U, as shown in Fig. 2, one from the BaF_2 clusters and the other from the $LaBr_3$:Ce detectors. The BaF_2 spectrum here has been scaled according to the BaF_2 spectrum from the ^{238}U(n, f) measurement (cf. Fig. 1), hence both have the same height relative to each other just like it is shown in Ref. [10]. The $LaBr_3$:Ce spectrum was adjusted to the BaF_2 spectrum by requiring the same integral, i.e. multiplicity, within the overlap region 0.5 MeV $< E_\gamma <$ 2.0 MeV. Finally, from the combined spectrum, PFGS characteristics for the fast-neutron induced fission of ^{235}U were deduced like described e.g. in Ref. [11]. The results are given in Table 1, together with corresponding values from another recent experiment on the same fissioning system, but induced by thermal neutrons [16]. The spectrum obtained from the latter measurement is also depicted in Fig. 2 for comparison. Apparently, all three spectra agree rather well, as the numbers do in Table 1 – within overlapping uncertainties. For the fast-neutron induced PFGS results, the uncertainties include statistical ones as well as those from the adjustments made and the transformation function. The results will be discussed below.

3 Comparing apples and oranges

The coincident measurement of photons and fission fragments gives rise to events of different physical origin [17], of which here only prompt fission γ rays are of interest. These events are

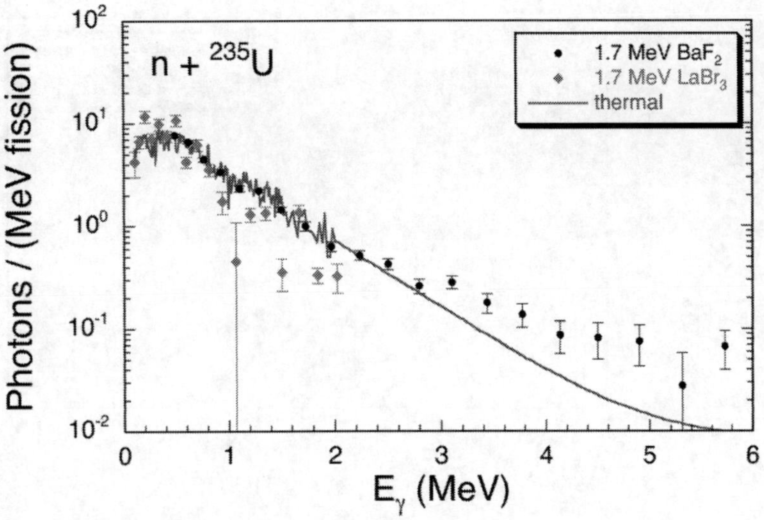

Figure 2. Prompt-fission γ-ray emission spectra for the reaction n + ^{235}U at $\overline{E}_n$ = 1.7 MeV obtained with the BaF_2 clusters [10] and with the $LaBr_3$:Ce detectors (this work). For comparison a recently measured spectrum from thermal neutrons on ^{235}U [16] is shown too.

easily distinguished by time-of-flight (TOF) measurements and identified as peak in a corresponding TOF distribution. Selecting γ rays inside this peak ensures that neutron-induced events are excluded, whereas any background is strongly suppressed by the coincidence condition between photons and fission fragments. During the last 60 years of experimental studies of prompt fission γ-ray emission, different detector systems, electronics, measurement and data analysis techniques were employed, which may cause difficulties when comparing the obtained results in terms of prompt fission γ-ray spectra (PFGS) characteristics. Hence, comparing experimental results is sometimes like "comparing apples and oranges". In previous works, the influence of the high energy range (i.e. above 6 - 10 MeV) of the detected PFGS [7] as well as low-energy photons (below 500 keV) [5] has been assessed already.

In order to illustrate the impact of the width of the timing window on PFGS characteristics, we refer here to previously unpublished results from a measurement on ^{252}Cf(sf). In this case both experimental setup and data treatment correspond to what has been described in Ref. [4]. However, for this study different TOF windows were used to select prompt fission γ rays and the resulting measured spectra with the corresponding events were unfolded with the response function of the used 58 mm × 58 mm cylindrical $LaBr_3$:Ce scintillation detector (for details about the unfolding procedure we refer to Ref. [4]).

Figure 3 shows the measured TAC (= time-to-analogue converter) spectrum for γ rays detected in coincidence with fragments from the spontaneous fission of ^{252}Cf as full-drawn line, centered around zero, i.e. the instant of fission. The dotted regions correspond to the

Table 1. Summary of PFGS characteristics from the reaction ^{235}U(n, f) at different incident neutron energies for the range 0.1 MeV < E_γ < 6.0 MeV.

Detector	$\overline{E}_n$ (MeV)	$\overline{M}_\gamma$	$\overline{E}_{\gamma,tot}$ (MeV)	$\overline{\epsilon}_\gamma$ (MeV)	Reference
$LaBr_3$:Ce/BaF_2	1.7	7.11 ± 0.44	5.51 ± 0.46	0.77 ± 0.08	this work
$LaBr_3$:Ce	thermal	7.38 ± 0.18	6.37 ± 0.20	0.86 ± 0.03	[16]

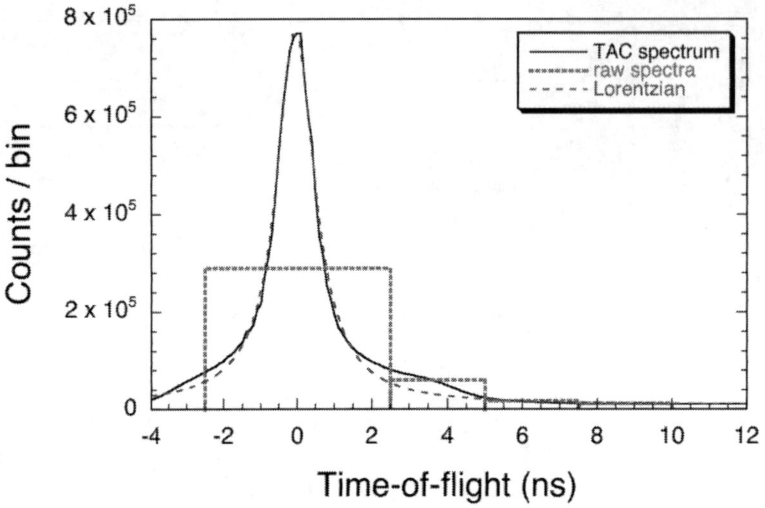

Figure 3. Time-of-flight distribution (denoted by TAC spectrum) of γ rays from a coincidence measurement with fission fragments from ^{252}Cf(sf), together with the sum of counts from the corresponding raw spectra averaged over the respective time regions. The Lorentzian is the result of a fit to the measured TAC spectrum.

number of γ rays in the measured raw spectra for each of the chosen time-of-flight regions, distributed over the respective bin. The limits are –2.5, 2.5, 5, 7.5 and 10 ns, respectively. Background contributions are very small and not depicted. Integrating over the regions separated by the dotted lines and over the TAC spectra within the same boundaries gives the same results. In addition, a Lorentzian is plotted as the fit result to the measured TAC spectrum; this approximation will be used later to parametrize the prompt γ-ray timing peak.

From the different prompt fission γ-ray emission spectra and according to the time limits depicted above, average multiplicities, mean energies per photon and average total γ-ray energies released in fission were then extracted. The obtained results are depicted as full dots in Fig. 4 as function of the upper time limit of the different timing windows, while the lower limit in all cases was –2.5 ns. The left-hand side shows average prompt fission γ-ray multiplicities $\overline{M}_\gamma$, while the average γ-ray energies $\overline{\epsilon}_\gamma$ are shown to the right. The error bars contain both statistical and systematic uncertainties. It is obvious that the average multiplicity increases with increasing upper time limit, which is not a surprise, since a wider time window implies more events to be included. In contrast, the mean energy per photon is practically constant over the studied time range; the mean value is indicated by a dashed vertical line, which agrees well with previously published values [4, 5]. This behavior is not really surprising either, since by widening the prompt time window the contributions from the additional γ rays become smaller and smaller. This implies that the measured energy spectra for each of these time windows should have the same shape, which by unfolding of the response function must lead to emission spectra, which have the same shape, too. Hence, the number of events within a certain window of the time-of-flight distribution should be proportional to the average multiplicity obtained by means of a proper unfolding of the measured energy spectrum of the corresponding γ rays. This is demonstrated by integrating over the TAC spectrum from a certain lower time limit to any upper time limit, which - normalized to the actually determined multiplicities - results in the line in the left part of Fig. 4, in perfect agreement with the experimental values. Finally, since $\overline{\epsilon}_\gamma$ is practically

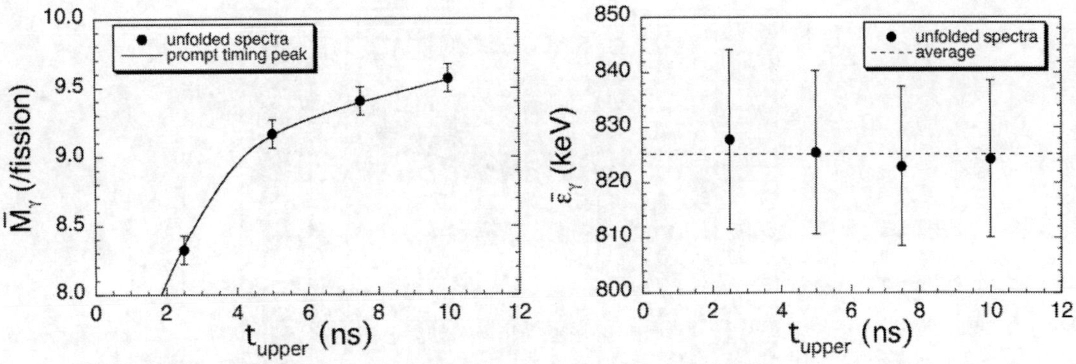

Figure 4. Average prompt fission γ-ray multiplicity $\overline{M}_\gamma$ and mean energy per photon $\overline{\epsilon}_\gamma$ from ^{252}Cf(sf), obtained for different prompt timing windows. The lower time limit is in all cases -2.5 ns, while the upper limits correspond to the regions in Fig. 3. See text for more details.

constant, we may conclude that the average total γ-ray energy released in fission, $\overline{E}_{\gamma,tot} = \overline{M}_\gamma \times \overline{\epsilon}_\gamma$, exhibits the same behavior as $\overline{M}_\gamma$.

In order to generalize time-of-flight distributions of prompt γ rays, we suggest to apply an approximation by a Lorentzian, as depicted in Fig. 3. This way it is possible to estimate multiplicities for any prompt timing window (albeit within reasonable limits) by simply integrating over an analytical mathematical function. With "reasonable limits" we refer here to the question, how late after fission a γ ray may still be considered prompt. This question was discussed in Ref. [18] and will be addressed by us elsewhere. Anyway, since the width of the prompt peak is a measure for the experimental coincidence timing resolution, it varies usually from experiment to experiment depending on employed detectors and other electronics. Hence, by expressing all times in units of the observed timing resolution, i.e. the full width at half maximum (FWHM) of the prompt peak, one becomes independent from particular experimental conditions.

Figure 5 shows results of those integrations in form of a nomogram. A relative multiplicity is plotted as a function of the upper integration limit t_{upper} and parametrized by the lower limit t_{lower}. All times are given in units of the timing resolution Δt_{FWHM}, in order to be applied to any experimental condition. The full drawn lines connect points of same value for t_{lower} and varying t_{upper}, describing asymmetric prompt timing windows. Results for symmetric timing distributions, i.e. for t_{upper} = -t_{lower}, are depicted as dotted line. Here, a value of one for the relative multiplicity has been chosen for a symmetric window which is 5 × Δt_{FWHM} wide. If a prompt fission γ-ray multiplicity has been determined for a certain timing window and a corresponding value for another experiment and/or timing window is to be inferred, the nomogram in Fig. 5 may be used as follows: for a performed multiplicity measurement with known timing resolution and prompt window as reference, the relative multiplicity is read depending on the limits of the chosen time window. Then the same is done for experimental conditions for which the result needs to be estimated. Finally, the new absolute multiplicity value is obtained by multiplying the reference value with the ratio of both relative multiplicities. As an example data from Ref. [11] is added in Fig. 5 as black dots. They were obtained for ^{252}Cf(sf) with a coincidence timing resolution of Δt_{FWHM} = 1.25 ns and for different symmetric prompt windows. According to the convention chosen here, a relative multiplicity of one is associated with a total witdth of 6.25 ns. Obviously, the data points follow very well the predicted dependence shown by the dotted line, thus indicating that the suggested

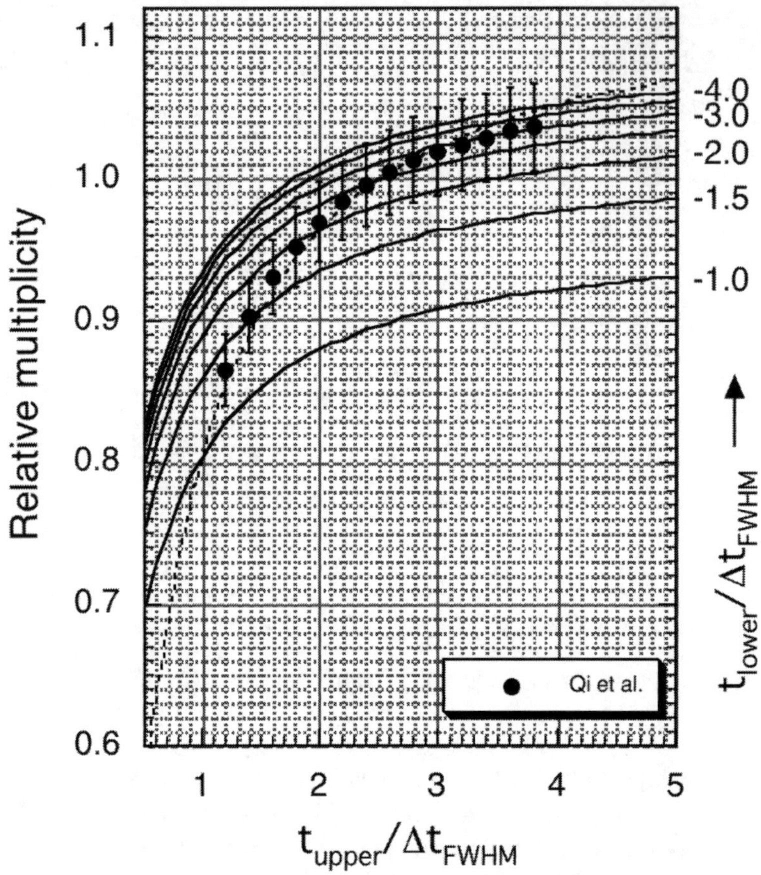

Figure 5. Nomogram to illustrate the calculated dependence of the (relative) prompt γ-ray multiplicity from the width of the chosen time window around the instant of fission. Upper and lower time limits are given in units of coincidence timing resolution Δt_{FWHM} (see text for details). Experimental data from Qi *et al.* [11] are shown for comparison.

procedure is indeed a useful tool to compare prompt fission γ-ray multiplicities from different experiments, i.e. to compare "apples and oranges".

4 Discussion and outlook

In the present work we have reported on two aspects of the measurement of fission fragment de-excitation by prompt γ-ray emission.

For the system n + ^{235}U we have presented the very first results for fast-neutron induced prompt fission γ-ray characteristics, which was achieved by combining results from a previously performed comparative measurement [15] and a recent study on n + ^{238}U [11] at almost identical incident neutron energy. The deduced PFGS characteristics are in rather good agreement with a recent measurement with thermal neutrons [16] – within the uncertainties. Still, all values are somewhat lower for fast neutrons than for thermal ones, which is certainly opposite to our expectations [14], but this was already observed for the system n + ^{239}Pu [8, 11]. It has to be admitted that the statistics was quite low, but nevertheless, it is the published first PFGS results for fast neutrons on ^{235}U. Obviously, a new measurement with better statistics is highly recommended.

In detailed studies of the spontaneous fission of ^{252}Cf we investigated the dependence of PFGS characteristics from the width of the chosen prompt time window. Within the range ±10 ns we observed a with time increasing (but apparently saturating) average prompt γ-ray multiplicity – and consequently total γ-ray energy, since the average energy per photon appears to be constant. This behavior could be explained by integration over the prompt timing distribution within different limits. By expressing the time dependence in units of a respective experimental coincidence timing resolution, conclusions can be made independent from a particular experiment. Moreover, results from different measurements can be compared with each other, which has been illustrated graphically. In the next future, we will extend the investigated time scale by several orders of magnitude to address the questions of how late after the instant of fission emitted γ rays may still be considered prompt and how experimental findings compare to theoretical predictions.

One of the authors (A. O.) acknowledges the support from the Extreme Light Infrastructure Nuclear Physics (ELI-NP) Phase II, a project co-financed by the Romanian Government and the European Union through the European Regional Development Fund - the Competitiveness Operational Programme (1/07.07.2016, COP, ID 1334), with which this work had been finalized.

References

[1] R. Vandenbosch and J.R. Huizenga, Nuclear Fission, Academic Press New York and London (1973)

[2] C. Wagemans (editor), The Nuclear Fission Process, CRC Press Boca Raton (1991)

[3] S. Oberstedt, R. Billnert, F.-J. Hambsch, M. Lebois, A. Oberstedt, and J.N. Wilson, Eur. Phys. J. A **51**, 178 (2015)

[4] R. Billnert, F.-J. Hambsch, A. Oberstedt, and S. Oberstedt, Phys. Rev. C **87**, 024601 (2013)

[5] A. Oberstedt, R. Billnert, F.-J. Hambsch, and S. Oberstedt, Phys. Rev. C **92**, 014618 (2015)

[6] S. Oberstedt, A. Oberstedt, A. Gatera, A. Göök, F.-J. Hambsch, A. Moens, G. Sibbens, D. Vanleeuw, and M. Vidali, Phys. Rev. C **93**, 054603 (2016)

[7] A. Oberstedt, T. Belgya, R. Billnert, R. Borcea, T. Bryś, W. Geerts, A. Göök, F.-J. Hambsch, Z. Kis, T. Martinez, S. Oberstedt, L. Szentmiklosi, K. Takács, and M. Vidali, Phys. Rev. C **87**, 051602(R) (2013)

[8] A. Gatera, T. Belgya, A. Göök, F.-J. Hambsch, M. Lebois, A. Oberstedt, S. Oberstedt, L. Qi, L. Szentmiklósi, G. Sibbens, and M. Vidali, Phys. Rev. C **95**, 064609 (2017)

[9] S. Oberstedt, R. Billnert, T. Belgya, T. Bryś, W. Geerts, C. Guerrero, F.-J. Hambsch, Z. Kis, A. Moens, A. Oberstedt, G. Sibbens, L. Szentmiklosi, D. Vanleeuw, and M. Vidali, Phys. Rev. C **90**, 024618 (2014)

[10] M. Lebois, J.N. Wilson, P. Halipré, A. Oberstedt, S. Oberstedt, P. Marini, C. Schmitt, S. J. Rose, S. Siem, M. Fallot, A. Porta, and A.-A. Zakari, Phys. Rev. C **92**, 034618 (2015)

[11] L. Qi, M. Lebois, J.N. Wilson, A. Chatillon, S. Courtin, G. Fruet, G. Georgiev, D. G. Jenkins, B. Laurent, L. Le Meur, A. Maj, P. Marini, I. Matea, L. Morris, V. Nanal, P. Napiorkowski, A. Oberstedt, S. Oberstedt, C. Schmitt, O. Serot, M. Stanoiu, and B. Wasilewska, Phys. Rev. C **98**, 014612 (2018)

[12] J.-M. Laborie, R. Billnert, G. Bélier, A. Oberstedt, S. Oberstedt, and J. Taieb, Phys. Rev. C **98**, 054604 (2018)

[13] L. Qi, J.N. Wilson, M. Lebois, A. Al-Adili, A. Chatillon, D. Choudhury, A. Gatera, G. Georgiev, A. Göök, B. Laurent, A. Maj, I. Matea, A. Oberstedt, S. Oberstedt, S.J. Rose, C. Schmitt, B. Wasilewska and F. Zeiser, EPJ Web of Conf. **169**, 00018 (2018)

[14] A. Oberstedt, R. Billnert, and S. Oberstedt, Phys. Rev. C **96**, 034612 (2017)

[15] M. Lebois, J.N. Wilson, P. Halipré, B. Leniau, I. Matea, A. Oberstedt, S. Oberstedt, D. Verney, Nucl. Instr. Methods Phys. Res. A **735**, 145 (2014)

[16] A. Gatera, PhD thesis, Ghent University (2019)

[17] A. Oberstedt, S. Oberstedt, R. Billnert, W. Geerts, F.-J. Hambsch, J. Karlsson, Nucl. Instr. Methods Phys. Res. A **668**, 14 (2012)

[18] P. Talou, T. Kawano, I. Stetcu, J.P. Lestone, E. McKigney, and M. B. Chadwick, Phys. Rev. C **94**, 064613 (2016)

Energy sharing based on microscopic level densities

Jφrgen Randrup[1,2,*], *Martin* Albertsson[3], *Gillis* Carlsson[3], *Thomas* Dφssing[4], *Peter* Möller[5], and *Sven* Åberg[3]

[1]Lawrence Berkeley National Laboratory, Berkeley, California 94720, USA
[2]Yukawa Institute for Theoretical Physics, Kyoto 606-8502, Japan
[3]Mathematical Physics, Lund University, S-221 00 Lund, Sweden
[4]Niels Bohr Institute, DK-2100 Copenhagen φ, Denmark
[5]Moller Scientific Graphics and Computing, Los Alamos, New Mexico 87545, USA

Abstract. The transformation of a moderately excited heavy nucleus into two excited fission fragments is modeled as a strongly damped evolution of the nuclear shape. The resulting Brownian motion in the multi-dimensional deformation space is guided by the shape-dependent level density which has been calculated microscopically for each of nearly ten million shapes (given in the three-quadratic-surfaces parametrization) by using a previously developed combinatorial method that employs the same single-particle levels as those used for the calculation of the pairing and shell contributions to the five-dimensional macroscopic-microscopic potential-energy surface.

The stochastic shape evolution is followed until a small critical neck radius is reached, at which point the mass, charge, and shape of the two proto-fragments are extracted. The available excitation energy is divided statistically on the basis of the microscopic level densities associated with the two distorted fragments. Specific fragment structure features may cause the distribution of the energy disvision to deviate significantly from expectations based on a Fermi-gas level density.

After their formation at scission, the initially distorted fragments are being accelerated by their mutual Coulomb repulsion as their shapes relax to their equilibrium forms. The associated distortion energy is converted to additional excitation energy in the fully ac- celerated fragments. These subsequently undergo sequential neutron evaporation which is calculated using again the appropriate microscopic level densities. The resulting de- pendence of the mean neutron multiplicity on the fragment mass, $\overline{\nu}(A)$, as well as the dependence of $\overline{\nu}(A)$ on the initial excitation energy of the fissioning compound nucleus, exhibit features that are similar to the experimentally observed behavior, suggesting that the microscopic energy sharing mechanism plays an important role in low-energy fission.

*e-mail: JRandrup@LBL.gov

1 Introduction

We illustrate here how shape-dependent microscopic level densities may be used to refine the theoret- ical modeling of nuclear fission. Our focus is on fission of moderately excited compound nuclei, such a thermal neutron-induced fission of actinides.

2 Microscopic shape-dependent level density

In Ref. [1] a method was developed for microscopic calculations of level densities for deformed nuclei and it has been adapted to the fission process [2]. For a specified shape χ, the single-particle levels for protons and neutrons needed for the combinatorial calculation of the level density are obtained by solving the Schrödinger equation in the associated folded-Yukawa potential. These are the same levels as those previously used in Ref. [3] to calculate the microscopic shell and pairing energies in the construction of the deformation energy surface, thus guaranteeing consistency of the treatment. The corresponding local many-body vacuum state $|0;\chi\rangle$ has N neutrons and Z protons filling the lowest single-particle states and the uncorrelated excited states consist of all multiple particle-hole excitations,

$$|i;\chi\rangle = \prod_{n\geq 1} a^{\dagger}_{\nu_n^{(i)}} a_{\mu_n^{(i)}} |0;\chi\rangle , \tag{1}$$

For each many-body state $|i;\chi\rangle$, blocked BCS calculations for neutrons and protons separately provide the state-dependent pairing gaps, $\Delta_i^n(\chi)$ and $\Delta_i^{\rm p}(\chi)$, respectively, and the energy of the correlated intrinsic many-body state, $E_i(\chi) = E_i^n(\chi) + E_i^{\rm p}(\chi)$.

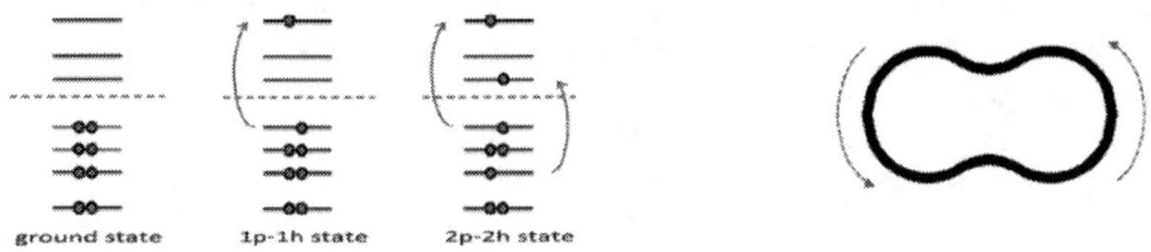

Figure 1. Illustration of the calculation of the microscopic level density for a specified shape, $\rho(E, I, \chi)$. For each many-particle-many-hole excitation, a BCS pairing calculculation is carried out for neutrons and protons separately, yielding the associated correlated energy of the intrinsic state, $E_i^n(\chi) + E_i^{\rm p}(\chi)$ (*left panel*); each such intrinsic state forms the head of a rotational band contributing a rotational energy, $E_i^{\rm rot}(I;\chi)$ (*right panel*).

For each such intrinsic state, the angular momentum along the nuclear symmetry axis is denoted by K and it is assume that it forms a rotational band head. The resulting total angular momenta I may then take on the values $I = K$, $(K + 1)$, $(K + 2)$, . . . and the corresponding rotational energies of the band members are

$$E_i^{\rm rot}(I;\chi) = \left[I(I+1) - K_i(\chi)^2\right] / \left[2\mathcal{J}_{\perp}(\chi, \Delta_i^{\rm n}(\chi), \Delta_i^{\rm p}(\chi))\right] . \tag{2}$$

The moment of inertia $\zeta_{\perp}$ is approximated by the moment of inertia of a rigid body with the shape χ, modified by the calculated pairing gaps for the state [4].

In Ref. [1] it was found that the enhancements arising from collective vibrational modes are unim- portant and they are therefore ignored in the present fission applications. The total energy of a state is then given by

$$E_i(I,\chi) = E_i^{\rm n}(\chi) + E_i^{\rm p}(\chi) + E_i^{\rm rot}(I;\chi) . \tag{3}$$

In terms of these energies, the level density is obtained by binning,

$$\rho(E,I,\chi) = \frac{1}{\Delta E}\int_{E}^{E+\Delta E} \sum_{i} \delta(E' - E_i(I,\chi))\, dE' \,. \tag{4}$$

Thus, for each shape χ, the resulting states are binned according to their energy E_i and their total angular momentum I; the bin width was taken as $\Delta E = 200$ keV. The sensitivity of the results to the bin width has been tested and it was found that a doubling or tripling of ΔE has no discernable effect on the calculated observables.

3 Brownian shape evolution

The description of nuclear fission as a generalized Brownian motion builds on the assumption that the evolving nucleus can be characterized by its shape degrees of freedom χ. The associated shape param- eters are treated as classical variables that are coupled dissipatively to the remaining nuclear degrees of freedom which have a primarily microscopic character and are assumed to have a short relaxation time. The resulting large-amplitude collective motion then exhibits a strongly damped diffusive evo- lution that can be described as a random walk on the associated multi-dimensional potential-energy surface $U(\chi)$. When the values of the potential are available on a discrete lattice of shapes (as is the case for the 5D tabulations made in Ref. [5]), the Smoluchowski equation can be approximately simulated by means of a random walk on this lattice [6].

The rates of transition between two lattice sites χ and χ', $\nu(\chi \leftrightarrow \chi')$, and the corresponding local level densities, $\rho(\chi)$ and $\rho(\chi')$, must satisfy detailed balance, $\rho(\chi\rightarrow)\ \nu(\chi \rightarrow \chi') = \rho(\chi')\ \nu(\chi' \rightarrow \chi)$. A particularly simple way of ensuring dertailed balance is presented by the Metropolis procedure [7] according to which a proposed shape change from χ to χ' is accepted unconditionally if $\rho(\chi') > \rho(\chi)$ ("downhill"), whereas it is accepted only with the probabilty $\rho(\chi')/\rho(\chi)$ otherwise ("uphill").
In Ref. [6] and all subsequent appli,cations until recently [2], a simplified Fermi-gas level-density expression was used, $\tilde{\rho}_E(\chi) \approx \exp[2\sqrt{a_A(E - U(\chi))}]$ with $a_A \approx A/(10\ \text{MeV})$. Because the level density then depends on the shape χ only through the local excitation energy $E^*(\chi) = E - U(\chi)$, the ratio between neighboring level densities can be approximated by means of a logarithmic expansion,

$$\frac{\tilde{\rho}(\chi')}{\tilde{\rho}(\chi)} \approx \exp\left[-\frac{\partial \ln \tilde{\rho}(E^*)}{\partial E^*}\frac{\partial U(\chi)}{\partial \chi}\Delta\chi\right] \approx e^{-\Delta U/T} \,, \tag{5}$$

where $\Delta U = U(\chi') - U(\chi)$ is the change in the potential energy associated with the proposed shape change $\chi \rightarrow \chi' = \chi + \Delta\chi$ and $T(\chi) = 1/[\partial \ln \tilde{\rho}(E^*)/\partial E^*]$ is the local nuclear temperature. In the present treatment, we follow Ref. [2] and employ the microscopic level densities (see Sect. 2) to evaluate the required ratios $\rho(\chi')/\rho(\chi)$. The use of microscopic level densities is particularly well- suited for describing the dependence of the shape evolution on the total excitation energy of the fissioning system because the microscopic pairing and shell effects automatically subside as the energy is raised, without the need for any adjustable parameters.

4 Scission

We focus here on fission of excited ^{236}U. An ensemble of scission configurations is generated by performing one million Metropolis walks, starting from well inside the outer fission barrier

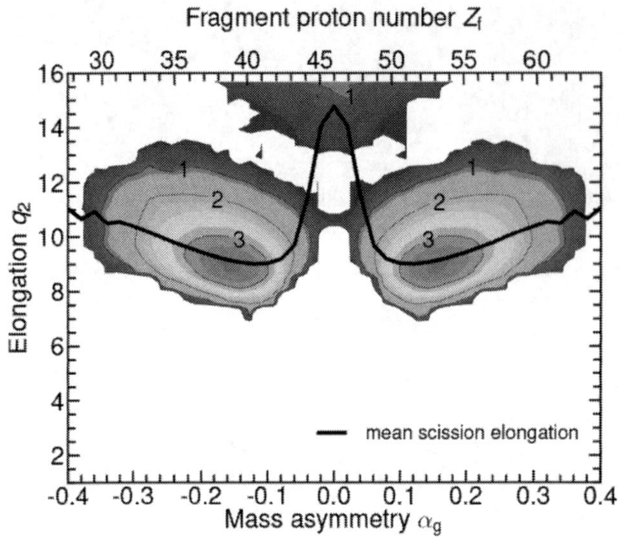

Figure 2. Logarithmic contour plot of the location of the end shapes projected onto the plane of mass asymmetry α_g (corresponding the lattice index $M = -33 : 33$) and relative quadrupole moment q_2 (corresponding the lattice index $I = 1 : 45$ for ^{235}U(n_{th},f); the fragment charge numbers $Z_f = 46(1 + \alpha_g)$ are shown on the upper axis (from Ref. [2]). The distribution has a dominant asymmetric component and a relatively small symmetric component at significantly larger elongations. Because the lattice employed in [2] does not extend sufficiently far in q_2 to fully encompass the symmetric component it has been extended upwards to $I = 53$ for the present study.

and continuing until the neck has shrunk to a specified critical value, $c_0 = 1.5$ fm. Figure 2 illustrates the distribution of these end shapes.

Each end shape is assumed to be abruptly transformed into two proto-fission fragments whose shapes are obtained by completing the two spheroidal shape segements, as illustrated in Fig. 3. (An overall rescaling is subsequently performed to conserve the total volume.)

The proto-fragments are generally distorted relative to their equilibrium shapes to which they relax as they move apart due to their mutual Coulomb repulsion. The associated distortion energies,

$$E_i^{\text{dist}} \approx M_i(\varepsilon^{\text{sc}}) - M_i(\varepsilon^{\text{gs}}),\ i = \text{H, L}, \tag{6}$$

are assumed to be converted into additioanl statistical excitation energy in the respective fragments before the neutron evaporation commences.

Because the collective kinetic energy associated with the shape evolution is assumed to be neg- ligible prior to scission, we take the available excitation energy at scission as the difference between the total energy, E, and the potential energy of the end configuration,

$$E_{\text{sc}}^* \ = \ E - U(q_2^{\text{sc}}, c^{\text{sc}}, \varepsilon_1^{\text{sc}}, \varepsilon_2^{\text{sc}}, \alpha_g^{\text{sc}})\,. \tag{7}$$

The key assumption in the present study is that this energy is divided among the two proto-fragments statistically based on the microscopic level densities associated with their distorted shapes. Thus the excitation energy in the heavy fragment is governed by the following probability distribution,

$$P(E_H^*; E_{\text{sc}}^*) \ \sim \ \rho_H(E_H^*; \varepsilon_H^{\text{sc}})\rho_L(E_{\text{sc}}^* - E_H^*; \varepsilon_L^{\text{sc}})\,. \tag{8}$$

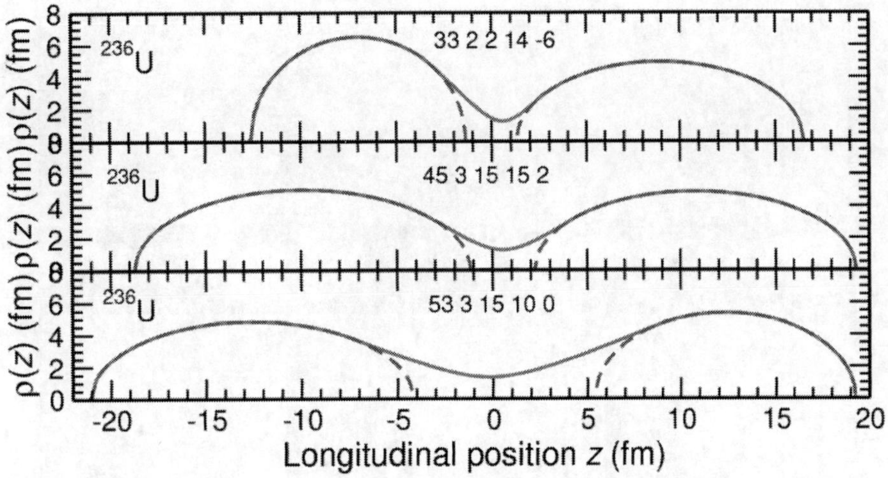

Figure 3. Three examples of the scission procedure: For each end shape (characterized by having a neck radius c equal to the specified critical value, c_0 = 1.5 fm), each of the two spheroidal endcaps of the 3QS shape is completed and the resulting two-spheroid shape is renormalized to conserve volume. The top panel shows an asymmetric end shape (which is fairly compact), whereas the two lower panels show end shapes that are nearly mass symmetric (and rather elongated). The shape lattice indices (*I, J, K, L, M*) are indicated for each shape.

Figure 4 shows the energy distribution $P(E_{\rm H}; E^*_{\rm sc})$ at three different values of the total available energy $E^*_{\rm sc}$ for two different mass divisions having $(A_{\rm H} : A_{\rm L}) = (130 : 106)$ and $(152 : 84)$. These two divisions contribute to the yields at the inner and outer wings of the double-humped mass distribution, respectively, and the fragment deformations considered are typical of those divisions.

For the case shown in the top panel of Fig. 4, the heavy fragment, ^{130}Sn, is very close to being doubly magic and it therefore has a spherical ground-state shape, $\varepsilon^{\rm gs}_{\rm H} = 0$, while the light partner fragment, ^{106}Mo, has a well-deformed prolate ground-state shape, $\varepsilon^{\rm gs}_{\rm L} = 0.33$. The fragment deformations at scission are $\varepsilon^{\rm sc}_{\rm H} = -0.10$ and $\varepsilon^{\rm sc}_{\rm L} = 0.30$ which both deviate only slightly from the respective ground-state deformations. The near magicity of the heavy fragment (with a shell correction energy of –10.2 MeV) causes the level density to remain very small up to excitation energies of 20 MeV. Conversely, the shell correction energy of the light fragment is +0.35 MeV and its level density is considerably larger than that of the heavy partner in that energy range. As a consequence, the major part of the energy goes to the light fragment, the most probable division being about 2 : 8 MeV. This is very different from the Fermi-gas scenario in which the most probable division is 5.5 : 4.5 MeV.

The opposite appears when the two fragments differ more in size, as illustrated in the bottom panel of Fig. 4. Here the microscopic energy-partition distribution function strongly favors the heavy fragment, ^{152}Nd, relative to the light fragment, ^{84}Ge. In this case, the typical scission deformation of the heavy fragment is considerably smaller, $\varepsilon^{\rm sc}_{\rm H} = 0.10$, than its ground-state deformation, $\varepsilon^{\rm gs}_{\rm H} = 0.24$. Therefore the heavy fragment has a large single-particle level density and, consequently, it has a large positive shell correction, ≈+6.1 MeV (as compared to ≈ 6.9 MeV for the ground-state shape) and a particularly high level density. On the other hand, the light fragment, ^{84}Ge, has a neutron number that is close to being magic, $N_{\rm L} = 52$, so its level density is low. As a result, the heavy fragment is strongly favored in the energy division, even up to quite high energies, as clearly seen in Fig. 4 (d)–(f).

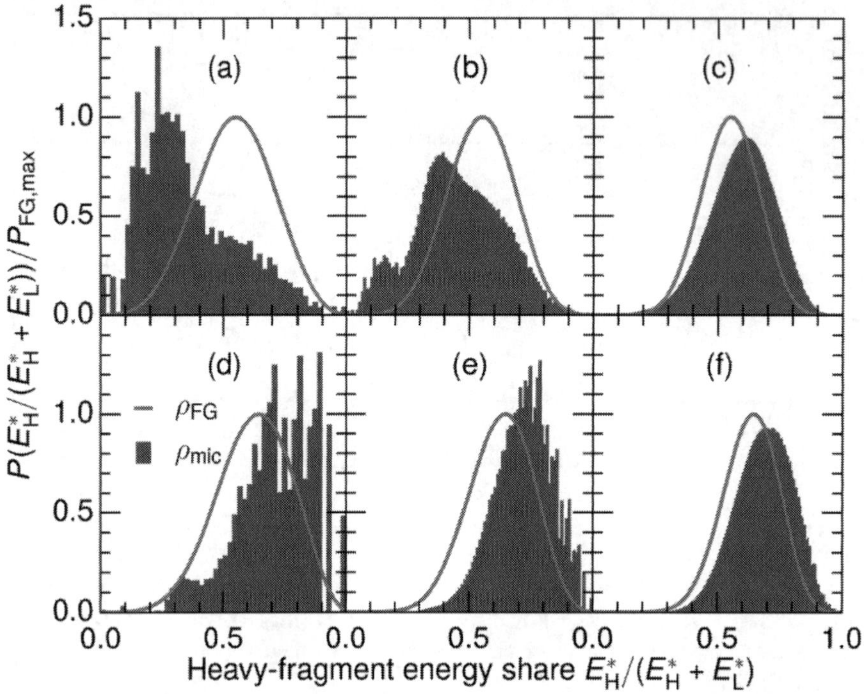

Figure 4. For ^{235}U(n,f) is shown the distribution function $P(E^*_H; E^*_{sc})$ giving the probability that the heavy primary fission fragment acquires the excitation energy E^*_H when the total excitation energy available at scission is E^*_{sc}, for two different divisions, either 130+106: $(N, Z, \varepsilon)_H = (80, 50, -0.1)$ and $(N, Z, \varepsilon)_L = (64, 42, +0.3)$ (*top panels*) or 152+ 84: $(N, Z, \varepsilon)_H = (92, 60, +0.1)$ and $(N, Z, \varepsilon)_L = (52, 32, +0.1)$ (*bottom panels*), and three different values of the total available energy at scission: $E^*_{sc} = 10$ (left column), 20 (center column), 40 (right column) MeV. In each panel, the distributions obtained from the shape-dependent microscopic level densities $\rho(E;\chi)$ (blue histograms) and those obtained with the simplified Fermi-gas level density $\rho_{FG}(E^*)$ (red curves) have been scaled with a common factor so that the maximum value of the Fermi-gas result is unity.

As the available energy E^*_{sc} is increased, the microscopic energy partition distribution, $P(E^*_H; E^*_{sc})$, approaches the macroscopic form obtained with the simplified Fermi-gas level density, $\tilde{\rho}_E(\chi)$, and the structure effects in the energy partition distribution gradually subside.

For each scission configuration obtained at the end of the Metropolis walk, the excitation ener- gies of the proto-fragments are sampled from the appropriate microscopic partition distribution (8) illustrated in Fig. 4. Figure 5 shows the mean excitation and distortion energies, $\bar{E}_{sc}(A)$ and $\bar{E}_{dist}(A)$, together with their sum, the resulting total excitation energy of the primary fission fragment.

The mean fragment excitation energy at scission has a pronounced structure that may be qual- itatively understood from the energy partition distribution functions of the two examples discussed above. The local minimum slightly below $A \approx 130$ and the local maximum around ≈ 106 result from the favoring of the light fragment in the energy sharing, illustrated in Fig. 4 (a) – (c), while the pronounced maximum at $A \approx 150$ and the relatively low values in the $A \approx 84$ region result from the favoring of the heavy fragment, illustrated in Fig. 4 (d) – (f).

Due to overall energy conservation, the energetics of the fragments is coupled to their kinetic en- ergies (which are not discussed here due to space limitations). For the asymmetric fission component (see Fig. 2), the mass dependence of the total fragment kinetic energies,

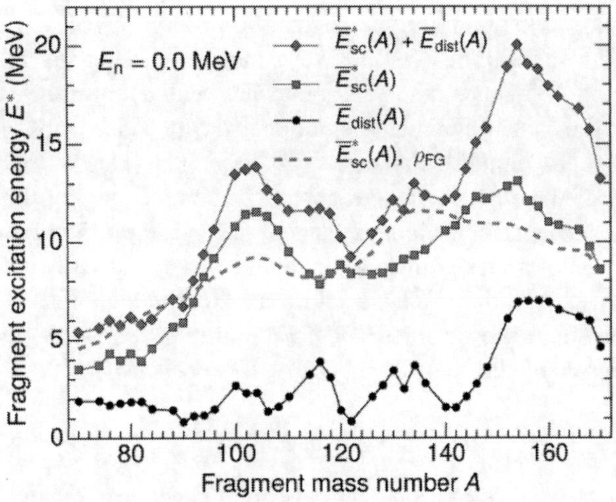

Figure 5. As functions of the fragment mass number A are shown the mean fragment distortion energy $E_{dist}(A)$ (black dots), the mean excitation energy at scission, E^*_{sc} (A) (red squares), and the sum, $E_{dist}(A)$ + E^*_{sc} (A) (green diamonds). as extracted from an ensemble of 10^6 scission configurations. For reference is also shown the value of$E^*_{sc}(A)$ that would be obtained with the simple Fermi-gas level density $\rho_{FG}(E^*)$ (red dashes).

TKE(A), generally agrees very well with the experimental data. But for the symmetric component TKE comes out too high which leads to too low values of E^*_{sc}. This problem is currently under investigation.

After a fragment has been fully accelerated and its shape has relaxed to its ground-state form, it disposes of its excitation energy by neutron evaporation (see Appendix A) and, on a longer time scale, by radiation of photons. The dependence of the mean neutron multiplicity on the fragment mass, $\bar{\nu}(A)$, may be used to test the calculated energy partitioning.

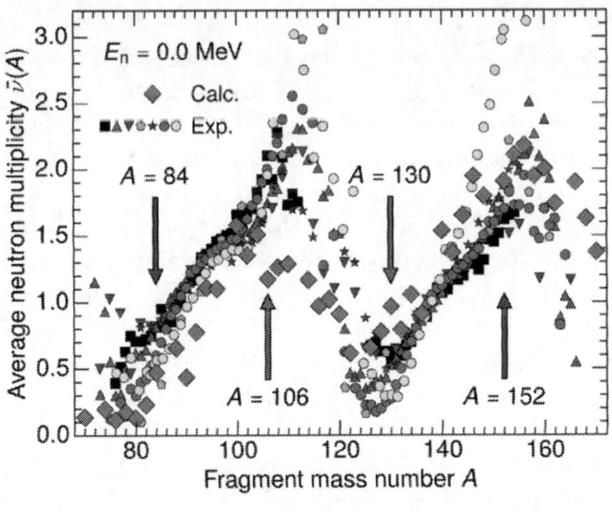

Figure 6. The mean neutron multiplicity, $\bar{\nu}$, as a function of the mass number of the primary fission fragment, A, for ^{235}U(n$_{th}$,f). The calculated results (blue diamonds) are shown together with available experimental data: Nishio *et al.* [8] (black squares), Apalin *et al.* [9] (yellow circles), Vorobyev *et al.* [10] (green triangles), Batenkov *et al.* [11] (orange diamonds), Boldeman *et al.* [12] (purple stars), Maslin *et al.* [13] (brown triangles), Göök *et al.* [14] (red circles). The red/blue arrows point to the mass divisions examined in Fig. 4.

Figure 6 shows the calculated mean neutron multiplicity $\bar{\nu}(A)$ together with data from a variety of experiments. The sawtooth appearance of the data is reasonably well reproduced by the calculation. It arises from a combined effect of the behavior of the neutron separation energy, $S_n(A)$, which displays a jump near A = 132 due to the closed shells at Z = 50 and N

= 82, and the behavior of the total fragment excitation energy $E^*_{\rm sc}(A) + E^*_{\rm dist}(A)$ displayed in Fig. 5. The significant shortfall of $\bar{\nu}(A)$ in the symmetric region arises in large part from the fact that the employed 3QS shape family is somewhat inadquate for the very elongated scission configurations of the mass-symmetric component (which also leads to too high fragment kinetic energies for near-symmetric divisions).

We have also studied how $\bar{\nu}(A)$ changes as the excitation energy of the fissioning nucleus is in- creased. The sawtooth behaviour is then weakened due to the reduction of the shell corrections near $A = 130$ which significantly increases the level density in the heavy fragment and hence $\bar{\nu}_{\rm H}$. Such an evolution is also seen in the experimental data. The calculated $\bar{\nu}(A)$ gradually approaches the form obtained if the excitation energy were divided according to the masses, as suggested by the simplified Fermi-gas level density.

5 Conclusion

In order to elucidate how the excitation energy is divided between the two fission fragments, we have augmented the recently developed level-density guided Metropolis shape evolution treatment [2] with shape-dependent microscopic level densities for the proto-fragments which are distorted relative to their equilibrium shapes. The available energy is partitioned statistically according to the corresponding microscopic level densities which take account of the structure effects in the proto-fragments. After separation, the distortion energy is converted into additional excitation before the sequential neutron evaporation commences. The dependence of the resulting mean neutron multiplicity on the mass number of the primary fragment agrees reasonably well with the data. In particular, the sawtooth appearance can be understood to result in part from shell-structure effects in the level densities as well as from structure in the multi-dimensional potential-energy surface.

This report is based in part on the collaborative work reported in Ref. [16]. It was supported by the Swedish Natural Science Research Council (BGC and SÅ) and by the Office of Nuclear Physics in the U.S. Department of Energy's Office of Science under Contract No. DE-AC02-05CH11231 (JR).

Appendix A Neutron evaporation

The fragment-mass dependence of the average neutron multiplicity, $\bar{\upsilon}$, presents a convenient observable that is sensitive to the division of the excitation energy bewteen the primary fragments. We have therefore simulated the evaporation cascades using the simple Weisskopf-Ewing treatment in which the neutron spectrum is assumed to be of the form, $dN/d\varepsilon \sim \sqrt{\varepsilon}\rho(E^*_{\rm max} - \varepsilon)\sqrt{\varepsilon}d\varepsilon$. The last factor comes from the neutron phase space d^3p and the level density refers to the daughter at the excitation $E^*_{\rm max} - \varepsilon$ where $E^*_{\rm max} = E^* - S_{\rm n}$ is the maximum possible excitation energy in that fragment after the evaporation (corresponding to the emission of a neutron with vanishing kinetic energy, $\varepsilon = 0$); E^* is the excitation energy of the mother fragment and $S_{\rm n}$ is its neutron separation energy. The first factor represents the transmission coefficient which is assumed to be proportional to the transverse speed of the neutron. In the present work, we use the microscopic level density for the evaporation daughter nucleus, evaluated for its ground-state shape $\varepsilon_{\rm gs}$, and we repeat the evaporation procedure as long as it is energetically posisble to emit a neutron, *i.e.* as long as E^* exceeds $S_{\rm n}$, because the competiton from photon emission has an insignificant effect on $\bar{\upsilon}$.

Often an analytical form of the level density is employed (for example, in FREYA, a simplified Fermi-gas from is used, $\rho(E^*) \sim \exp(2\sqrt{a_A E^*})$). Then $\ln\rho(E^*)$ may be expanded around $E^* = E^*_{\rm max}$,

$$\ln\rho(E^*_{\rm max} - \varepsilon) \approx \ln\rho(E^*_{\rm max}) - \varepsilon\left[d\ln\rho(E^*)/dE^*\right]_{E^*=E^*_{\rm max}} = \ln\rho(E^*_{\rm max}) - \varepsilon/T_{\rm max}, \quad (9)$$

where $T_{\max}$ is the maximum temperature in the daughter, leading to $dN/d\varepsilon \sim \varepsilon \exp(-\varepsilon/T_{\max})$ d_{ε}. For the present explorative study, this simple treatment would suffice.

References

[1] H. Uhrenholt, S. Åberg, A. Dobrowolski, T. Døssing, T. Ichikawa, and P. Möller, Nucl. Phys. A **913**, 127 (2013)

[2] D. Ward, B. G. Carlsson, T. Døssing, P. Möller, J. Randrup, and S. Åberg, Phys. Rev. C **95**, 024618 (2017)

[3] P. Möller and J. Randrup, Phys. Rev. C **91**, 044316 (2015)

[4] R. Bengtsson and S. Åberg, Phys. Lett. B **172**, 277 (1986)

[5] P. Möller, A.J. Sierk, T. Ichikawa, A. Iwamoto, R. Bengtsson, H. Uhrenholt, and S. Åberg, Phys. Rev. C **79**, 064304 (2009)

[6] J. Randrup and P. Möller, Phys. Rev. Lett. **106**, 132503 (2011)

[7] N. Metropolis, A.W. Rosenbluth, M.N. Rosenbluth, A.H. Teller, and E. Teller, J. Chem. Phys. **21**, 1087 (1953)

[8] K. Nishio Y. Nakagome, H. Yamamoto, and I. Kimura, Nucl. Phys. A **632**, 540 (1998)

[9] V.F. Apalin, Y. N. Gritsyuk, I.E. Kutikov, V.I. Lebedev, and L. A. Mikaelian, Nucl. Phys. A **71**, 553 (1965)

[10] A.S. Vorobyev *et al.*, EPJ Web Conf. **8**, 3004 (2010)

[11] O.I. Batenkov *et al.*, AIP Conf. Proc. **769**, 625 (2005)

[12] J.W. Boldeman, A.R. de L. Musgrove, and R. L. Walsh, Aust. J. Phys. **24**, 821 (1971)

[13] E.E. Maslin, A.L. Rodgers, and W.G.F. Core, Phys. Rev. **164**, 1520 (1967)

[14] A. Göök, F.J. Hambsch, and S. Oberstedt, EPJ Web Conf. **169**, 4 (2018)

[15] R. Müller, A.A. Naqvi, F. Käppeler, and F. Dickmann, Phys. Rev. C **29**, 885 (1984)

[16] M. Albertsson, B.G. Carlsson, T. Døssing, P. Möller, J. Randrup, S. Åberg, arXiv:1811.02283

EPJ Web of Conferences **256**, 00014 (2021) https://doi.org/10.1051/epjconf/202125600014
Theory-5

TANGRA multidetector systems for investigation of neutron-nuclear reactions at the JINR Frank Laboratory of Neutron Physics

Ivan Ruskov[1,2,*], *Yury* Kopach[1], *Vyacheslav* Bystritsky[1], *Vadim* Skoy[1], *Dimitar* Grozdanov[1,2], *Nikita* Fedorov[1], *Tatyana* Tretyakova[3], *Fuad* Aliev[4], *Constantin* Hramco[5], *Vyacheslav* Slepnev[1], *Nikolay* Zamyatin[1], *Aman* Gandhi[6], *Dongming* Wang[7], *Ajay* Kumar[6], *Evgeni* Zubarev[8], *Evgeny* Bogolubov[8], *Yuri* Barmakov[8], and *TANGRA* collaboration

[1]Joint Institute for Nuclear Research (JINR), Joliot-Curie str., 6, 141980 Dubna, Moscow region, Russia

[2]Institute for Nuclear Research and Nuclear Energy (INRNE) of Bulgarian Academy of Sciences (BAS), Tzarigradsko Chaussee Blvd. 72, 1784 Sofia, Bulgaria

[3]Skobeltsyn Institute of Nuclear Physics (SINP), Lomonosov Moscow States University, Lenin Hills 1, 119991 Moscow, Russia

[4]Azerbaijan Institute of Geology and Geophysics of Azerbaijan National Academy of Sciences, H. Javid Av. 119, Baku AZ1143, Azerbaijan

[5]Institute of Chemistry of the Academy of Sciences of Republic of Moldova, Academic str. 3, MD-2028, Chisinau, Republic of Moldova

[6]Department of Physics, Banaras Hindu University, Varanasi – 221005, India

[7]School of Energy and Power Engineering, Xi'an Jiaotong University, Xianning West Road, Xi'an, Shaanxi, 710049, P.R. China

[8]All-Russia Research Institute of Automation (VNIIA), P.O. Box 918, 101000 Moscow, Russia

Abstract. In the framework of TANGRA-project at the Frank Laboratory of Neutron Physics of the Joint Institute for Nuclear research in Dubna (Russia), two experimental setups (Fig. 1) have been designed and tested for investigation of 14-MeV neutron-induced nuclear reactions on a number of important for nuclear science and engineering isotopes. As a source of 14-MeV "tagged" neutrons we are using the VNIIA ING-27 steady-state portable neutron generator with embedded in its vacuum tube 64-pixel charge-particle detector. The "Romashka" system is an array of up-to 24 hexagonal NaI(Tl)-crystal scintillation probes, while the "Romasha" array consists of 18 cylindrical BGO-crystal detectors of neutrons and gamma-rays. In addition to these detectors there is a HPGe gamma-ray spectrometer and a number of Stilbene detectors that can be added for high-resolution gamma-ray spectrometry and neutron-gamma detection. The main characteristics of the neutron-induced nuclear reaction products can be investigated by commissioning the detectors in suitable for these experiments' geometries. Both setups can be used for doing basic and applied scientific research, because they permit simultaneously to measure the energy, angle and multiplicity distributions of gamma-rays and neutrons, produced in the competitive neutron-induced nuclear reactions (n, n'γ), (n,2n), (n, xnγ) and (n, f) in pure or complex substances.

*e-mail: ivan.n.ruskov@gmail.com;ivan@inrne.bas.bg

1 Introduction

Our society and the modern technologies now and in future will need much more energy that in the past. To respond to this demand, different types of energy production sources are utilized or are in the stage of development (solar, wind, hydro, fission, fusion). Between them the most effective sources of clean energy still are the nuclear power plants. To solve the problems with the utilization of the burned fuel and radioactive wastes from Generation-I-III nuclear power reactors, Generation-IV fast neutron reactors are under development. The conceptual design, construction and more strict requirements for protecting the environmental need more precise data about the interaction of neutrons with atomic nuclei of nuclear fuel and the construction materials. The libraries with experimental and evaluated nuclear data, particularly on ^{238}U, need to be enhanced with more precise and confident information about the characteristics (i.e. total and partial, differential, cross-sections) of different type of fast neutron induced reactions (elastic and nonelastic neutron scattering, fission, etc.), and of the products (mass-energy, multiplicity and angular distributions) from them [1, 2]. The gamma-rays producing in the (n, xnγ) and (n, fission) are of a particular ifnterest from fundamental, practical and methodological points of view. They are helping to investigate the nuclear structure of the atomic nuclei as well as the neutron induced processes themselves, performing theoretical calculations, model simulations and/or experiments. Gamma-rays from (n, n$'\gamma$) and (n, f) reactions with fast neutrons scientifically contribute to the radiation heating and damage of the construction and radiation shielding materials. The results of calculations and simulations need to be validated in the experiments that are, as a rule, quite complicated, resource and time consuming. Because the gamma-rays are following the disintegration of any excited nucleus of the neutron irradiated samples and of the environmental materials, the experimental setups consist of neutron and/or gamma detectors shielded or not depends on the setup and the reaction(s) under investigation. In the case of fast neutron induced nuclear reaction many possibilities (channels) are possible (open), i.e. (n, xnγ), (n,p), (n, α), (n, f), depends on the energy of the neutrons and the mass of the nuclei, and the use of multidetector systems is preferable, because in one measurement one can obtain, for example, the partial cross-section, angular and energy distributions of the reaction products. There are many and different types of multidetector systems for the investigation of the neutron induced reaction characteristics at neutron producing facilities all around the world.

At JINR FLNP in Dubna (Russia), in the frame of the international project TANGRA [3] we constructed 2 relatively low budget multidetector systems, namely, TANGRA-NaI(Tl) and TANGRA-BGO. TANGRA stands for TAgged Neutrons & Gamma Rays. That means that we are using tagged neutrons for investigating the characteristics of the gamma rays by arrays of NaI(Tl) and BGO crystal based probes. As a source of tagged neutrons, we are using the portable neutron generator ING-27 of VNIIA (Moscow) [4].

2 Experimental setups and their application

TANGRA Setups are multidetector, multipurpose, multifunctional, mobile systems, to study the characteristics of the products from the nuclear reactions induced by 14-MeV tagged neutrons [5–7]. TANGRA setups consist of a portable generator of "tagged" neutrons with energies of 14-MeV, ING-27, with or without an iron shield-collimator, 2D fast neutron beam profilometer, arrays of neutron-gamma detectors ("Romashka", "Romasha"), or single HPGe for high-resolution gamma-spectrometry and a computerized system for DAQ and analysis.

3 Portable Neutron Generator ING-27

The ING-27 DT Neutron Generator, manufactured by VNIIA (Moscow, Russia), utilizes the $^{3}H(d, n)^{4}He$ fusion reaction to produce 14-MeV neutrons. A beam of deuterons is accelerated up to ~120 keV (in lab-system) and bombards a tritiated Ti-target in the generator assembly (TiT). The generator has a rated 14-MeV neutron output of up to 10^8 n/s in 4π [4].

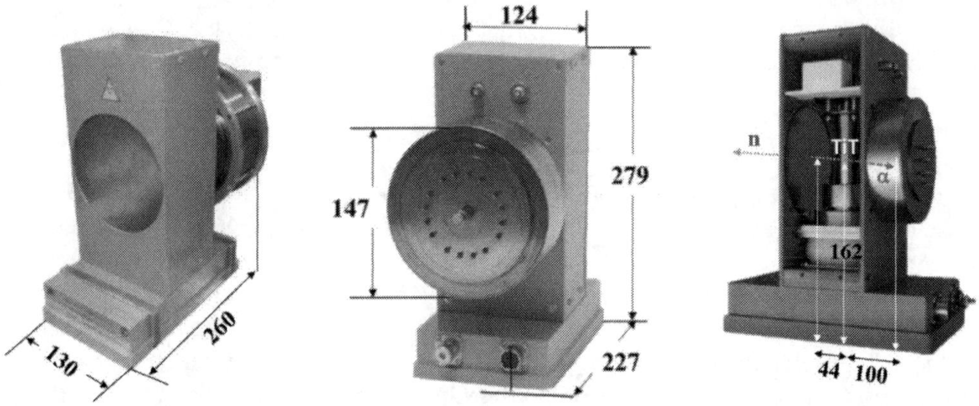

Figure 1. ING-27 dimensions (mm) and the direction of irradiation of DT-reaction fragments.

The α-particle detector incorporated into the ING-27 vacuum-chamber consists of 8 mutually perpendicular Si-strips on each side, mounted on a plastic frame and forming an 8×8-matrix of pixels with a size of 6×6 mm. The total sensitive area of the 64-element α-detector is 55×55 mm^2. The pixelated position-sensitive alpha-detector is located 100 mm away from the centre of the TiT-target (Fig. 1, right). It permits to "tag" the neutron, i.e. to determine the direction cone of the neutron travel (x, y-resolution) and to give the start-time for measuring the neutron time-of-flight spectra (z, depth resolution). Its front-end electronics, which consists of 16 charge-sensitive pre-amplifiers, is mounted on the back side of the neutron generator.

All the 3 experimental TANGRA setups include an ING-27 as a source of tagged 14-MeV neutrons. For registering the gamma-rays produced in the interaction of the neutrons with the nuclei of the investigated sample and the environment we are using 2 arrays from NaI(Tl) and BGO crystal based detectors. "Romashka" multidetector system consists of up to 22 hexagonal NaI(Tl) crystal-based probes, "Romasha" BGO multidetector system consists of up to 18 cylindrical BGO crystal-based probes [5–7]. The 3rd configuration at the moment has only one HPGe detector. "Romashka" NaI(Tl) radiation detectors can be arranged horizontally or vertically around the investigated sample at a desirable distance from it. The vertical configuration was chosen as having a bigger efficiency. "Romasha" BGO are arranged horizontally because of the crystal's geometry and dimensions. The main characteristics of the NaI(Tl), BGO and HPGe gamma-ray detectors are shown in Table 2. To measure the energy and angular distributions of gamma-rays form neutron induced nuclear reactions we are using NaI(Tl) and BGO detectors in configurations shown in Figure 2 (left and middle).

For obtaining the gamma-ray energy spectra with a high-resolution we are using a HPGe in a compact geometry, as shown in Fig. 2 (right). GEANT4 simulations of the experimental setups helped to obtain the optimal detector configurations, providing good efficiency for registration of gamma-rays and time-resolution.

Table 1. ING-27 characteristics.

Neutron generator	
TiT-to-front distance	44.0 ± 1.4 mm
TiT-to-α-detector distance	100 ± 2 mm
Power supply voltage	200 ± 5 V
Max Power Supply Current	300 ± 30 mV
Consumed Power	< 40 W
Weight: ING-27	7.5 ± 0.5 kg
Weight: Power Supply and Operation Unit	2.7 ± 0.3 kg
n-tube life-time	> 800 h
< ING-27 Duty time >	18 months
Double-side Si α-particle detector	
Number of pixels	64 (8×8 strips)
Single pixel area	6×6 mm2
Distance between strips	0.5 mm
Voltage bias	−250V DC
I_dark (250V DC, @20°C)	< 5 μA

Table 2. The main characteristics of gamma-neutron detecting systems.

Multi-detector array	NaI(Tl) "Romashka"	BGO "Romasha"	HPGe Ortec
Number of detectors	22	18	1
Type	Crystal hexagonal rod	Crystal cylindrical rod	GMX 30-83-PL-S,
Dimensions	(78 × 90 × 200 mm)	(⌀76 × 65 mm)	(⌀ 57.5 × 66.6 mm)
Photo-multiplier tube	Hamamatsu R1306	Hamamatsu R1307	
γ-Energy-resolution	~ 7.2% (0.662 MeV)	~ 10.4% (0.662 MeV)	~ 3.4% (0.662 MeV)
γ-Energy-resolution	~ 3.6% (4.437 MeV)	~ 4.0% (4.437 MeV)	~ 0.3% (4.437 MeV)
γ-Time-resolution	~ 3.8 ns (4.437 MeV)	~ 4.1 ns (4.437 MeV)	~ 6.1 ns (4.437MeV)

Figure 2. Experimental setups: TANGRA-NaI(Tl) (left), TABGRA-BGO (middle), TANGRA-HPGe (right).

4 Tagged fast neutron beam profilometer

To measure the tagged neutron beam profiles, we are using a 2D-detector of fast neutrons, made of 4 double-side stripped position-sensitive Si-detectors [8]. It is shown in Fig. 3 together with its main characteristics.

At the moment each 8 strips are grouped together, forming a 8×8-matrix with pixel size of ~1.5 × 1.5 cm.

The setup for measuring the tagged neutron beam profiles is shown in Fig. 6.

Number of strips	32 × 32
Strip's width	~1.8 mm
Detector's sizes	120 × 120 mm
Thickness	300 μm
n-efficiency	~0.8%

Figure 3. Fast neutron profilometer.

32-channel 100 MHz, 14 bit digitizer, in the form of 2 PCI-E cards; ~10^5 events/s for each input channel.

32-channel, 200 MHz, 11 bit digitizer, USB-3 connection; ~10^5 events/s for each input channel. This DAQ can work also with the HPGe detector.

Figure 4. AFI ADCM (left) and Digital Signal Recorder DSR-32 (right).

The neutron registration is done by detecting the charged particles (silicon recoil nuclei, protons, deuterons and alpha particles), formed as a result of the following neutron induced nuclear reactions: ^{28}Si(n,n′)^{28}Si, ^{28}Si(n,α)^{25}Mg, ^{28}Si(n,p)^{28}Al and ^{28}Si(n,d)^{27}Al, during the interaction of fast neutrons with the substance of the profilometer (silicon).

5 Data acquisition systems

For collecting the amplitude signals from pixelized α-sensor, γ-detector systems and n-profilometer, we are using 2 computerized data acquisition 32-channel ADC systems: AFI Electronics ADCM and Alpatov's Digital Signal Recorder DSR-32, shown in Fig. 4.

The digitized amplitude signals (waveforms) are stored in list-mode (one-by-one) on the computer hard-disk for further off-line analysis by user created C++ programs and CERN ROOT scripts.

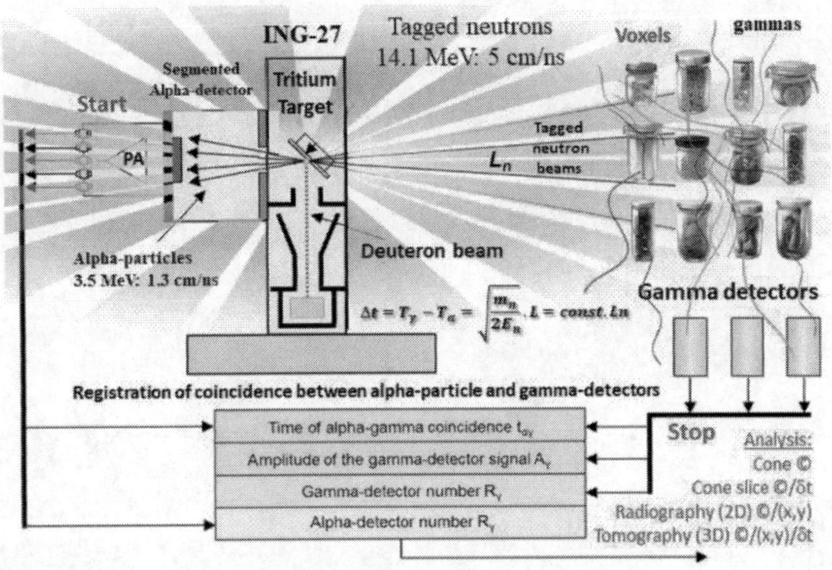

Figure 5. The scheme for investigating nuclear reactions by 14-MeV neutrons, using time-correlated associated α-particle technique.

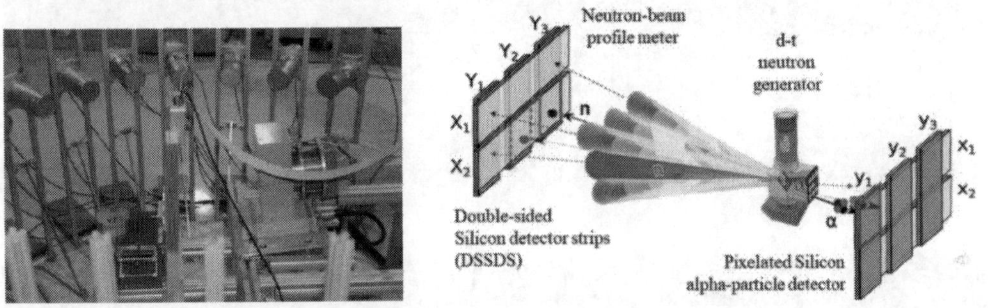

Figure 6. The photo of the setup (left) and the scheme for measuring the tagged neutron beam profiles (right).

6 Application of the tagged neutron method

For measuring the 14-MeV tagged neutron beam profiles and investigation the main characteristics of the neutron induced reactions we applied the Tagged Neutron Method (TNM), also known as Time-correlated Associated Particle (Imaging) Technique (Fig. 5).

The 14-MeV neutron is tagged in time and direction by detecting the associated α-particle, emitted in opposite direction (in CMS). The experimental setup for determination of the neutron flux in the ING-27 tagged neutron beams and their profiles with the n-profilometer by TNM is shown in Fig. 6.

To know the neutron flux and its profiles with a high precision is very important for absolute determination of the neutron induced reaction cross sections. The number of registered α-particles from DT-fusion reaction can be used as a monitor of neutrons irradiated from the TiT neutron producing target. The precision of monitoring depends on the detecting efficiency of the α-sensor, as well as from the number of lost (absorbed, scattered) α-particles prior to reach its sensitive area. We are using the n-profilometer for initial correct positioning

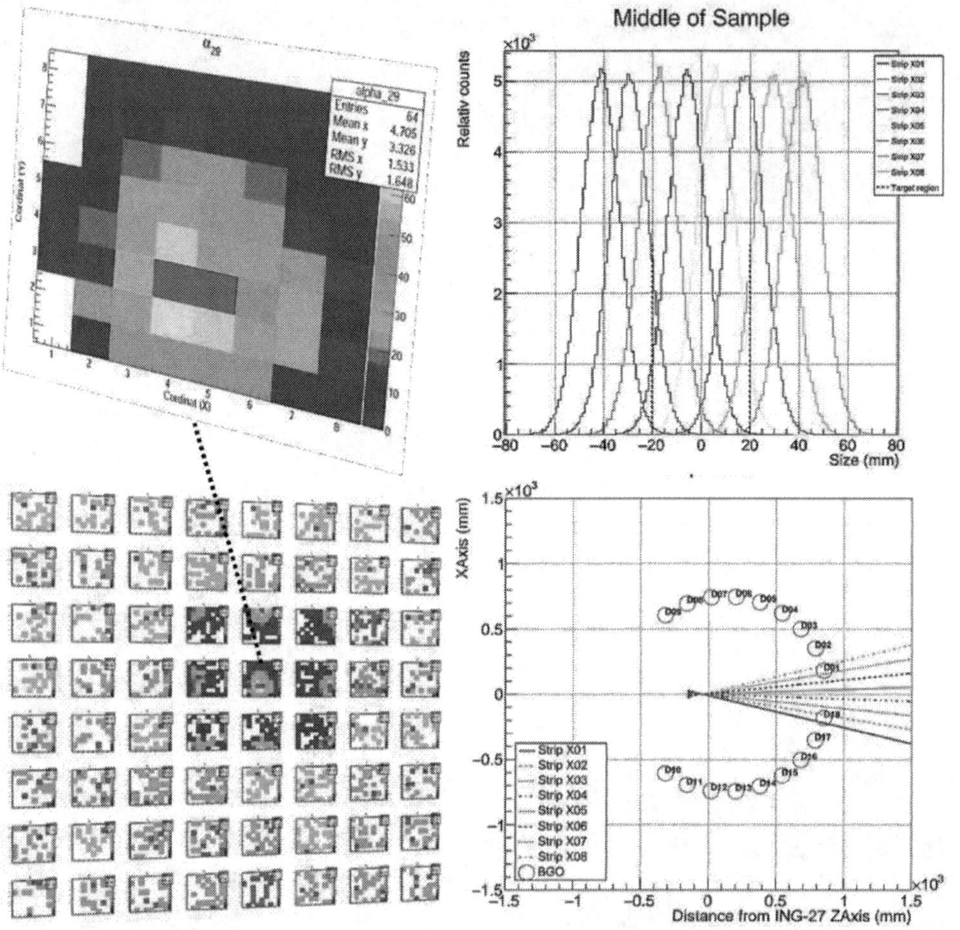

Figure 7. ING-27: 64 tagged neutron beams (down left), 2D-plot of one beam (up left), 8 tagged beams profiles (top right), azimuthal projection of 8 tagged beams cones (down right).

the target under investigation, to have as much as possible tagged beams hitting it, as well as detectors around it. An example of 2d-mapping of ING-27 tagged neutron beams, one of their profiles and the scheme of the experimental setup geometry with the beams' azimuthal projections are shown in Fig. 7.

Knowing the time "stamps" and the amplitudes of the signals it is possible not only to determine the neutron flux irradiating a certain voxel of the investigated object, but also to determine its location and elemental (isotope) composition. Situating the gamma (neutron) detectors around the sample in a way to avoid the multiple scattering effect, it is possible to measure the angular distribution of gamma-rays (and/or neutrons) and to obtain the angular correlations between them also, in addition to their energy spectra.

The associated particle technique for fast neutron interrogation of bulk samples is widely used for detection of explosives, buried land mines, chemical weapons, unexploded ordnance, drugs detection, in-vivo body composition, as well as of special nuclear materials (SNM).

By means of 2 multidetector systems, we made a feasibility study for measuring the angular distributions of gamma-rays from the inelastic scattering of 14.1-MeV neutrons on the nuclei of a number of light and middle weight isotopes, important for nuclear sciences, for which these data are not available or are known with a nonsufficient accuracy.

7 Test measurements with a bulk sample from ^{nat}U

To investigate the sensitivity of our TANGRA BGO and HPGe setups (Fig. 2, centre and right photos) for measuring the energy, anisotropy and multiplicity distributions of gamma-rays and the correlation of gamma-rays with the neutrons from the (n, xnγ), (n, f), (n, nf) reactions of 14-MeV neutrons on nuclei of some SNM, we made a short-term test with a bulk metal sample from natural uranium (^{nat}U).

The relative isotopic abundance of the sample is shown in Table 3.

Table 3. Relative Isotopic Abundance of the sample from ^{nat}U.

Isotope	By weight	By activity
U-238	99.282%	48.8%
U-235	0.712%	2.4%
U-234	0.006%	48.8%

Many factors may interfere with the prompt γ-ray emission from the inelastic scattering (n,n$'\gamma$) and fission (n,f) reactions, most of which are listed here: (1) delayed γ-rays emitted from the natU-sample; (2) prompt γ-rays from the T-target, ING-27 construction materials; (3) γ-rays from the detector and its surrounding material irradiated by scattered neutrons; (4) prompt γ-rays from the detector generated by penetrating neutrons; (5) prompt γ-rays from the Al-support of ING-27 and BGO-detectors; (6) spontaneous fission γ-rays from the sample.

The experimental data are in the process of analyzing and the results from it will be published elsewhere in a separate paper. Here, to highlight the problems one can have when is doing this type of measurements with SNM and to show the importance of using tagged neutrons and the TCAPM, we show only the energy spectra, obtained with TANGRA-HPGe setup (Fig. 2, right) and a rod-sample from ^{nat}U in the γ-ray energy interval from 60 keV up to 2 MeV. The measurements that were done are also indicated in Fig. 8. There are gamma-peaks from the ^{27}Al(n,n$'\gamma$) and peaks from the natural γ-radioactivity of the sample. The best way to minimize the influence of these parasitic sources of gammas is to apply the TCAPM, by which one can extract the gamma-ray energy spectra in different time-windows of the collected time-spectra, making the influence of the expected radiation background on the final results negligible.

8 Summary and outlook

In the framework of TANGRA project at JINR-FLNP were constructed 3 experimental setups for investigating the multiplicity, energy and angular distribution of gamma-rays and neutrons from induced by tagged 14-MeV neutrons (n, 2n), (n, xnγ) and (n,f) nuclear reactions. The experimental setups, consisted of ING-27 portable neutron generator and NaI(Tl), BGO, HPGe detectors, were successfully tested and used for measuring the angular distributions of gamma-rays from the inelastic scattering of 14-MeV neutrons with a number of materials used or supposed to be used as construction materials of today and future nuclear fission and fusion reactors. The further research activities are connected also with determination of neutron and gamma-ray emission from (n, n$'\gamma$) and (n, f) reactions of 14-MeV tagged neutrons with ^{238}U and other fertile and fissile heavy isotopes.

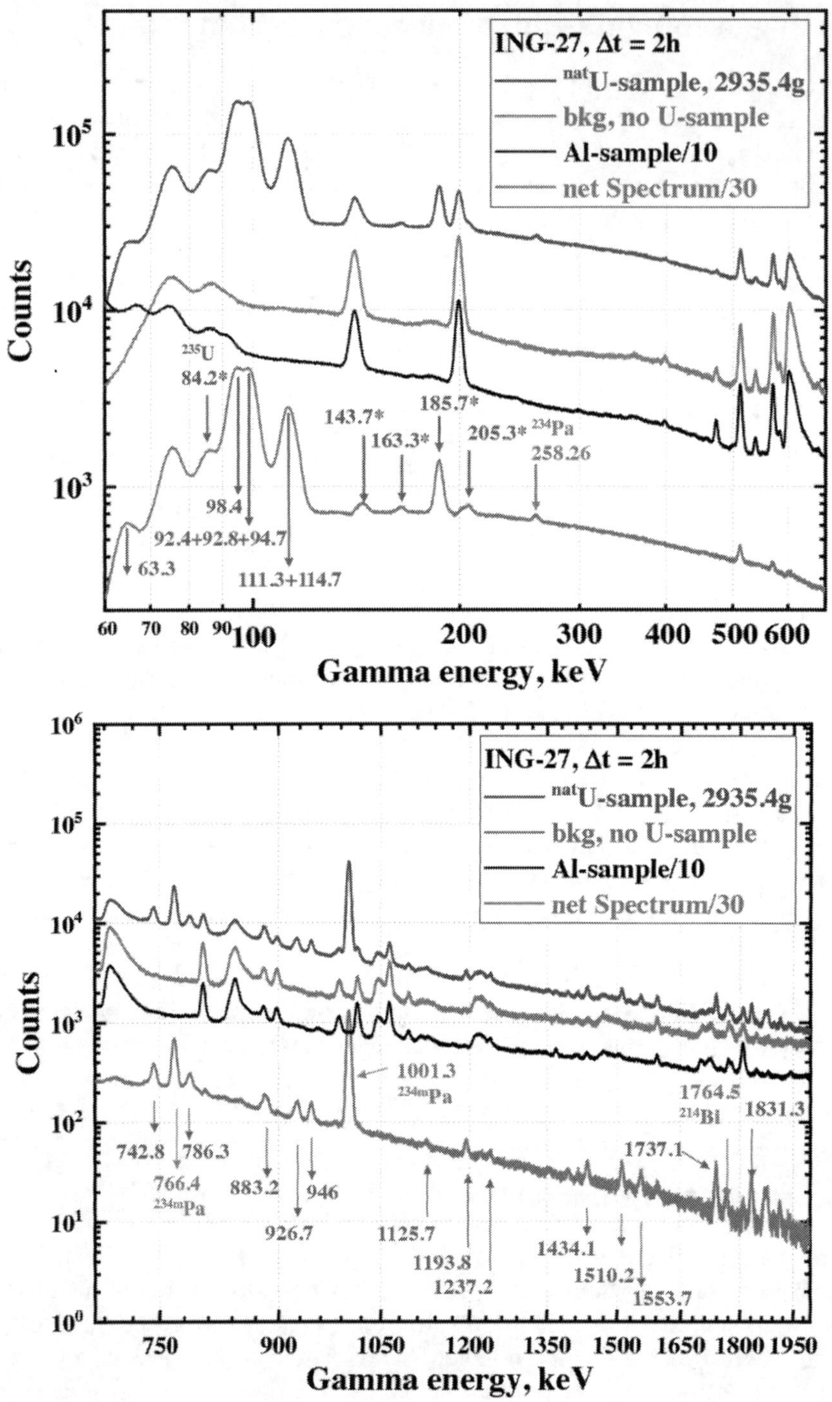

Figure 8. Energy spectra of the gamma-rays from ^{nat}U-sample.

References

[1] R. Capote, A. Trkov, M. Sin, M. Herman, A. Daskalakis, and Y. Danon, Physics of Neutron Interactions with ^{238}U: New Developments and Challenges, Nuclear Data Sheets **118** (2014) 26–31, doi:10.1016/j.nds.2014.04.003

[2] Roberto Capote and Andrej Trkov, Critical review of CIELO evaluations of n + ^{235}U, ^{238}U using differential experiments, EPJ Nuclear Sci. Technol. **4** (2018) 27, doi:10.1051/epjn/2018029

[3] Project TANGRA: "Design and development of the tagged neutron method for nuclear reaction studies and determination of the elemental structure of substances", JINR, Dubna, Russian Federation, http://flnph.jinr.ru/en/facilities/tangra-project

[4] E.P. Bogolyubov, A.V. Gavryuchenkov, M.D. Karetnikov, D.I. Yurkov, V.I. Ryzhkov, V.I. Zverev, Neutron generators and DAQ systems for tagged neutron technology, Proceedings of the XXVI International Symposium on Nucl. Electronics & Computing (NEC'2017), Becici, Budva, Montenegro, September 25–29, (2017) 176-181, http://ceur-ws.org/Vol-2023/176-181-paper-27.pdf, http://test.vniia.ru/eng/ng/element.html

[5] I.N. Ruskov, Yu.N. Kopatch, V.M. Bystritsky, V.R. Skoy, V.N. Shvetsov, F.-J. Hambsch, S. Oberstedt, R. Capote Noy, P.V. Sedyshev, D.N. Grozdanov, I.Zh. Ivanov, V.Yu. Aleksakhin, E.P. Bogolubov, Yu.N. Barmakov, S.V. Khabarov, A.V. Krasnoperov, A.R. Krylov, J. Obhodaš, L.B. Pikelner, V.L. Rapatskiy, A.V. Rogachev, Yu.N. Rogov, V.I. Ryzhkov, A.B. Sadovsky, R.A. Salmin, M.G. Sapozhnikov, V.M. Slepnev, D. Sudac, O.G. Tarasov, V. Valkovic, D.I. Yurkov, N.I. Zamyatin, Sh.S. Zeynalov, A.O. Zontikov, E.V. Zubarev. TANGRA-Setup for the Investigation of Nuclear Fission Induced by 14.1 MeV Neutrons, Physics Procedia **64** (2015) 163-170, doi:10.1016/j.phpro.2015.04.022

[6] I.N. Ruskov, Yu.N. Kopatch, V.M. Bystritsky, V.R. Skoy, V.N. Shvetsov, F.-J. Hambsch, S. Oberstedt, R. Capote Noy, and TANGRA collaboration, TANGRA-Setup for the Investigation of Nuclear Fission Induced by 14.1 MeV Neutrons, Scientific Workshop on Nuclear Fission dynamics and the Emission of Prompt Neutrons and Gamma Rays, THEORY-3, Physics Procedia **64** (2015) 163–170, doi:10.1016/j.phpro.2015.04.022

[7] Ivan Ruskov, Yury Kopatch, Vyacheslav Bystritsky, Vadim Skoy, Valery Shvetsov, Franz-Josef Hambsch, Stephan Oberstedt, Roberto Capote Noy, Dimitar Grozdanov, and TANGRA collaboration, TANGRA – an experimental setup for basic and applied nuclear research by means of 14.1 MeV neutrons, ND2016, EPJ Web of Conferences **146**, 03024 (2017), doi:10.1051/epjconf/201714603024.

[8] Zamyatin N.I., Bystritsky V.M., Kopach Y.N., Aliev F., Grozdanov D.N., Fedorov N.A., Hramko C., Ruskov I.N., Skoy V.R., Sleplev V.M., Wang D., Zubarev E.V., Neutron beam profilometer on the base of double-sided silicon strip detectors. Nuclear Instruments and Methods in Physics Research **A898** (2018) 46-52, doi:10.1016/j.nima.2018.04.031

Structural effects in the production of neutrons, gammas and anti-neutrinos in fission

K.-H. Schmidt[1,*], *M.* Estienne[1], *M.* Fallot[2], *B.* Jurado[3], and *C.* Schmitt[4]

[1]Rheinstr. 4, 64390 Erzhausen, Germany
[2]Subatech CNRS/IN2P3-Université de Nantes-IMTA, 4 rue Alfred Kastler, 44307 Nantes, France
[3]CENBG CNRS/IN2P3, Chemin du Solarium, B.P. 120, 33175 Gradignan, France
[4]IPHC CNRS/IN2P3, 23 rue du Loess, B.P. 28, 67037 Strasbourg, France

Abstract. Structural effects in the production of light particles in fission are investigated. Most of these effects can be traced back to pairing correlations and shell effects and their dependencies on the composition of the fissioning system and its excitation energy. It is shown that the GEF code is able to reproduce most of these features and to explain their origin on the basis of established properties of nuclear matter as well as concepts and laws of general validity. Predictions for systems with scarce or no experimental information can also be made.

1 Introduction

Light particles can be emitted during all processes connected to nuclear fission. It begins with the emission of nucleons, mostly neutrons, (before and after formation of a compound nucleus) from the excited system in the first minimum. This process is connected with multi-chance fission, and it causes discontinuities in almost all fission observables as a function of the initial excitation energy. Further on, neutrons can be emitted in the fission process before reaching the scission point. Eventually, scission neutrons (in analogy to light charged particles in ternary fission) may be emitted. After neck rupture, the fragments normally evaporate several neutrons, partly during the acceleration phase. Most gammas (statistical and non-statistical) are emitted from the fragments, predominantly from states below the angular-momentum-dependent neutron separation energy. The consecutive beta decay gives rise to the creation of electrons and anti-neutrinos and to the consecutive emission of beta-delayed gammas and neutrons. Structural effects are found in all of these processes.

2 Conventional modeling

Attempts for modeling the emission of prompt neutrons from fully accelerated fragments have a long tradition. The analytical Los-Alamos model [1] and later numerical computer codes (partly with the inclusion of gamma emission), e.g. [2–5], are in use to calculate the multiplicity distribution as well as the energy and angular distribution of the prompt

*e-mail: schmidt-erzhausen@t-online.de

neutrons as a statistical evaporation process. This is done by imposing measured or estimated fission-fragment Apre-TKEpre distributions, measured fragment-mass dependent mean prompt-neutron multiplicities and estimated Z distributions for a given Apre as well as estimated fragment angular-momentum distributions, which define the starting points of the evaporation cascades of the individual fragments. Models for the calculation of these starting conditions in case of scarce or missing experimental data, for example with a semi-empirical description [6], with microscopic calculations [6, 7] or with machine learning [8] are under development. Difficulties arise to obtain the necessary accuracy, to cover the multitude of the fission quantities, to assure the wide-spread correlations between them, and to make predictions for systems with scarce or no experimental information.

3 Features of a complete model

A complete model of light-particle emission in fission must consider the whole fission process, beginning from pre-equilibrium emission and evaporation of light particles before fission, the emission of mostly neutrons on the way to scission, eventually the production of scission neutrons, which is not finally confirmed, a model for the properties of the fission fragments, e.g. fragment yields, thermal excitation energy at scission and its division between the fragments, other contributions to the excitation energies of the final fragments, e.g. the release of deformation energy after scission, a model for the generation of fragment angular momentum and, as the last steps of prompt emission the evaporation of prompt neutrons and the emission of photons from the fragments, including the kinematical transformations from the fragment frame to the laboratory. Finally, the radioactive decay chains of the neutron-rich fragments must be considered.

The semi-empirical GEF model [9] fulfills most of these requirements. The flow diagram of GEF (Fig. 1) gives a graphical presentation of the described steps of the fission process and illustrates the complexity and the inherent correlations of the different fission quantities.

4 Some structural effects in comparison with the GEF model

This section is devoted to the description of some important basic ideas of the GEF model and to a comparison of specific results of the GEF code with experimental data, in view of light-particle production and gamma radiation in fission. For a comprehensive presentation of the GEF model, we refer to ref. [9].

4.1 Variation of fission yields with universal fragment shells

The regularities of fission channels as a function of the composition of the fissioning system and its excitation energy prove that the shell effects on the fission path evolve in a non-erratic way. The two-centre shell model provided the explanation that the shells slightly beyond the outer fission barrier are strongly determined by the fragment shells [10]. Therefore, only the macroscopic potential is characteristic for the fissioning system, while the shell effects are attributed to the fragments. This means that these shell effects are universal for all fissioning systems. The variation of the fission yields is caused by the superposition of the macroscopic and the microscopic contributions. This is illustrated in Fig. 2 on the example of the transition from predominantly symmetric to predominantly asymmetric fission around thorium.

In addition, observations [12] and theoretical considerations [13] suggest that the influence of proton shells is dominant. Thus, a proton shell at large deformation around Z = 55 (acting on the heavy fragment) seems to be responsible for the asymmetric fission in the actinides. However, also weaker shells in the light fragment and shells in the neutron subsystems of the light and the heavy fragment should have some additional influence.

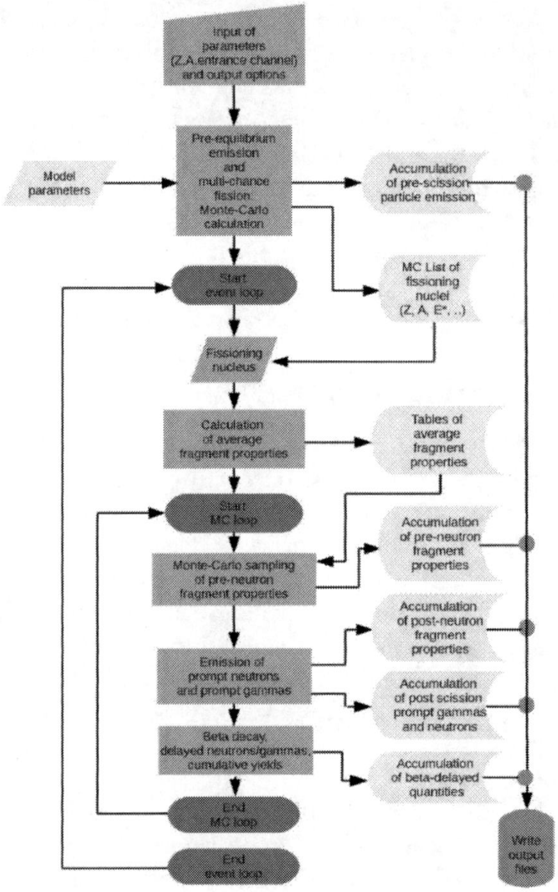

Figure 1. Flow diagram of the GEF code. (Figure modified from Ref. [9].)

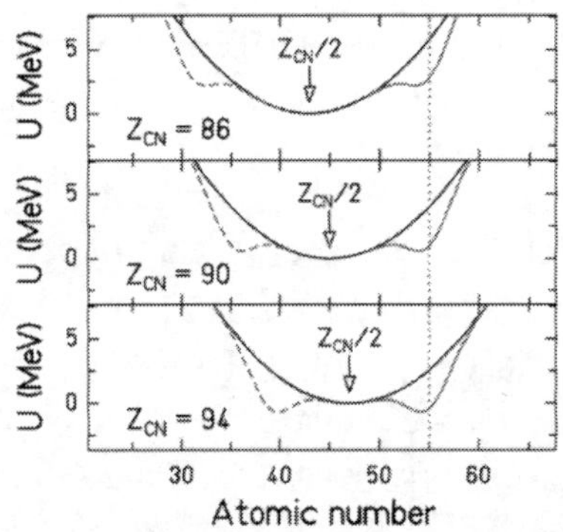

Figure 2. Schematic drawing of the superposition of the macroscopic potential (black) and a fragment shell at Z=55 (red) for different fissioning systems. (Figure from Ref. [9].)

4.2 Shell-determined fragment deformation at scission

It is a well known result of macroscopic-microscopic calculations (e.g. Ref. [11]) that shells at large deformation show a correlation between the particle number and the optimum deformation. This correlation can explain the saw-tooth behaviour of the mean prompt-neutron

 https://doi.org/10.1051/epjconf/202125600015

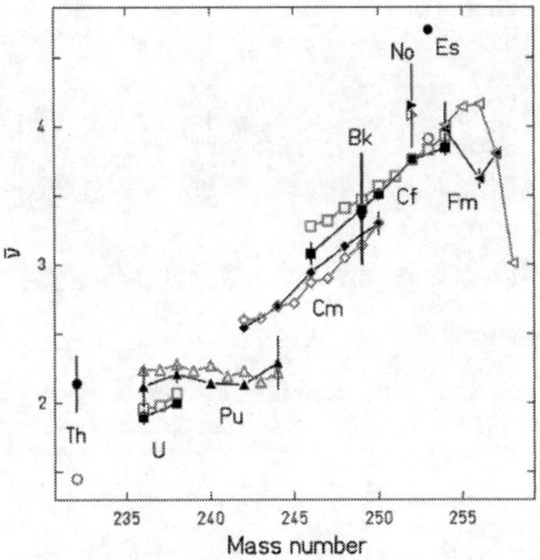

Figure 3. Measured prompt-neutron multiplicities in spontaneous fission (black symbols) in comparison with GEF calculations (red symbols). (Figure from Ref. [9].)

multiplicity in the fission of actinides. Together with the dominance of proton shells mentioned above, the contribution to the excitation energy of the fission fragments from the fragment deformation at scission is expected to be determined by the number of protons in the fragment.

4.3 Mean prompt-neutron multiplicities

With the ideas from Sects. 4.1 and 4.2 and a suitable unique set of model parameters, GEF is able to reproduce the mean prompt-neutron multiplicities as a function of the composition of the fissioning system to a great extent. Fig. 3 shows a comparison for spontaneous fission, where the structural effects are especially strong. There is a general increase of the neutron multiplicity with increasing mass along isotopic chains. The rather constant value for the Pu isotopes is explained by a relatively strong yield of the S1 fission channel that increases with neutron excess of the fissioning system. This feature is attributed to a shell in the light fragment near $Z = 38$, which is also assumed to be responsible for the double-humped fission of nuclei near ^{180}Hg [6]. The situation in the Fm isotopes is unclear, may be due to experimental uncertainties. Experimental verification of ^{256}Fm and additional data for other isotopes would be welcome.

4.4 Thermal energy at scission

Before scission, both nucleons and excitation energy can be exchanged between the nascent fragments through the neck. Commonly, the behaviour of such a di-nuclear system in fission or in binary heavy-ion reactions is assumed to be governed by statistical mechanics. e.g. by the tendency to acquire thermal equilibrium. In low-energy fission, the fissioning system is under the influence of pairing correlations, which means that the nuclear temperatures of the two nascent fragments do (almost) not depend on excitation energy. Due to the mass dependence of the nuclear temperature ($T \sim A^{-2/3}$), it is expected that the thermal energy available before scission is absorbed by the heavy fragment, because it has the lower temperature. This hypothesis has been tested by an experiment, where the incident energy in neutron-induced fission of ^{237}Np was increased from 0.8 MeV to 5.55 MeV [14].

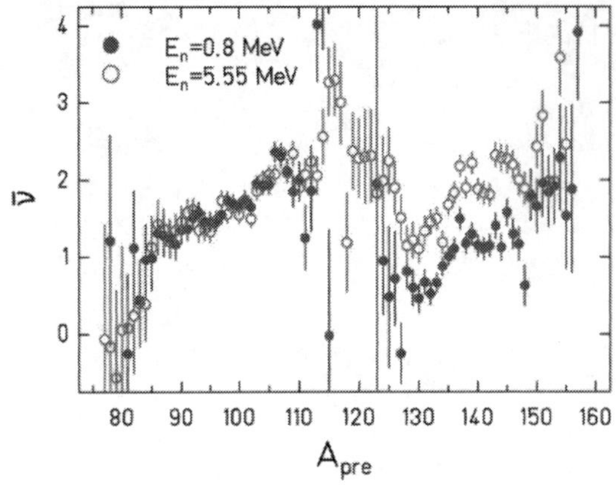

Figure 4. Mean multiplicity of prompt neutrons in the reaction ^{237}Np(n,f) with two different incident neutron energies as a function of fragment mass. Data from Ref. [14], figure from Ref. [9].

Indeed, the prompt-neutron multiplicity in the light fragment group did not change, while it increased all over the heavy-fragment range, see Fig. 4.

One could argue that the nuclei around the doubly-magic ^{132}Sn have strong shell effects, which increases the nuclear temperature. This could lead to an inversion of the energy-sorting process. However, in the relevant configuration at scission, this is only true for the fragments formed in the S1 fission channel. According to the GEF calculation shown in Fig. 5, the S1 fission channel has only a yield of 14.4%, and the yield is only comparable to the one of the S2 channel around A=107 and A=130. There is indeed an indication for an incomplete energy sorting in this region in Fig. 4, but the statistical significance is low. Thus, an eventually inverse energy sorting of fragments of the S1 channel is not in conflict with the experiment.

4.5 Prompt emission of gammas

GEF includes a competition between fission, light-particle emission and gamma emission from the first minimum of the initial system and between the emission of neutrons and gammas from the fragments. The gamma strength for the statistical emission of E1 radiation is given by the GDR as proposed in Ref. [15].

The prompt emission of non-statistical prompt gammas is governed by the complex spectroscopic nuclear properties. In the GEF code, these are described by an updated VMI (variable moment-of-inertia model). This makes calculations possible for nuclei with no or incomplete experimental spectroscopic information. The comparison with measured state energies shows rather good agreement, see Fig. 6.

The transition from rotational to vibrational structure in the vicinity of closed shells is well reproduced, although there are deviations for nuclei very near to closed shells.

Figure 7 shows rather good agreement for the low-energy part of the prompt-gamma spectrum for ^{239}Pu(n_{th},f). Note the higher resolution in the calculation!

4.6 Production of anti-neutrinos

The understanding of the anti-neutrino production in fission and the theoretical calculation of the anti-neutrino energy spectra in different, also future, types of fission reactors rely on the application of the summation method, where the individual contributions from the different

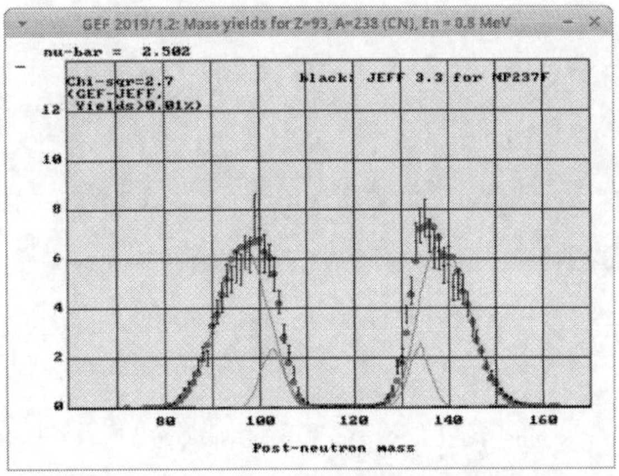

Figure 5. Mass dependent fission yields (after prompt-neutron emission) from GEF (red points) in comparison with JEFF 3.3 (black symbols with error bars). The green curves represent the contributions of the different fission channels. S2 is centered around A=96 and A=137, while S1 is centered around A=103 and A=133.

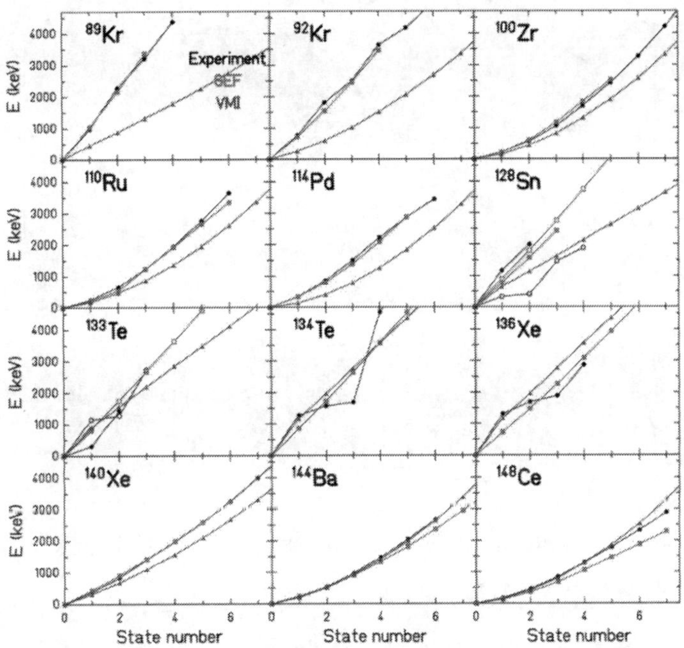

Figure 6. Experimental state energies (black dots) are compared with a rather recent VMI model [16] (blue) and the GEF model (red). Figure from [9].

radioactive nuclides that undergo a beta decay are estimated and summed up. Figure 8 shows the cumulative fission yields and the Q values of the beta decays for the system ^{235}U(n_{th},f), calculated with GEF by use of the JEFF 3.1.1 decay tables. The cumulative yields indicate the magnitude of the contributions of the different nuclei to the anti-neutrino spectrum, while the Q values give a hint on the maximum energy of these contributions. Apparently, the highest energies are expected for the most neutron-rich odd-*Z* light fission fragments.

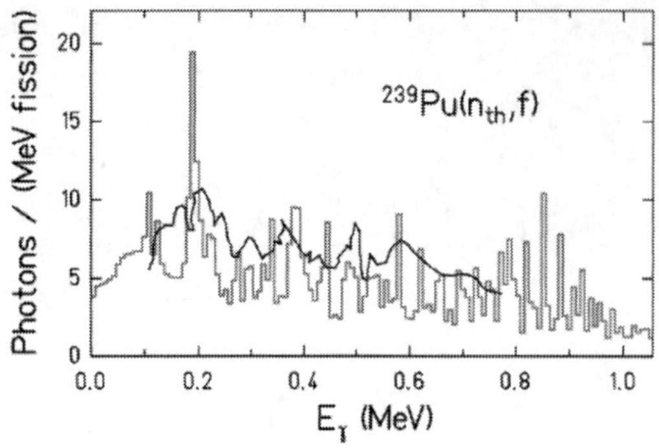

Figure 7. Lower part of the experimental prompt-gamma spectrum [17] (black line) of $^{239}Pu(n_{th},f)$ in comparison with the GEF result (pink histogram).

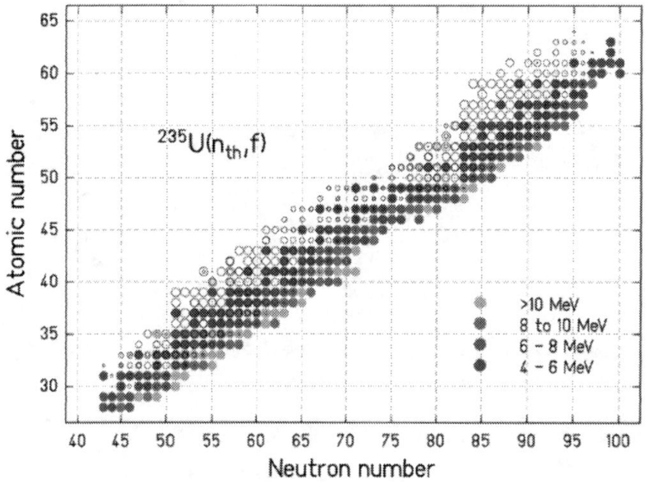

Figure 8. Cumulative yields (size of the black circles) and beta Q values (coloured dots) for the thermal-neutron-induced fission of ^{235}U from GEF. (Multiple black circles refer to the yields of isomers.)

Figure 9 shows the calculated anti-neutrino spectra from GEF for thermal-neutron-induced fission of different systems. The spectra show structures that are characteristic for the fissioning system and a tendency for the emission of high-energy neutrinos with increasing neutron excess. The accurate theoretical calculation of these spectra with their specific structures is a pre-requisite for an eventual remote monitoring of fission reactors.

4.7 Emission of delayed neutrons

The data on the evolution of delayed-neutron emission as a function of excitation energy are scarce. Nevertheless, there are clear indications for structural effects, see Fig. 10.

The GEF code reproduces the magnitude for the different systems rather well, while there is at least some qualitative similarities in the structures. The steep slope for even-Z systems around 5 MeV is well reproduced for $^{238}U(n,f)$ and not so well for other systems. It is created by and very sensitive to the energy dependence of the pairing gap. Another structure near 15 MeV is caused by multi-chance fission.

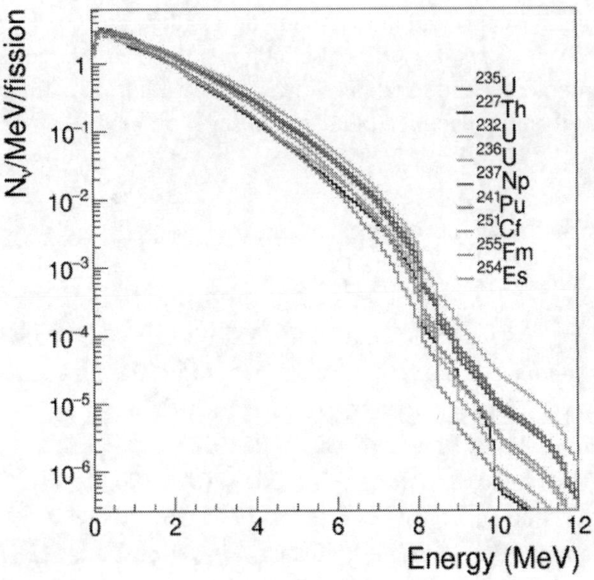

Figure 9. Calculated anti-neutrino spectra from GEF combined with the selection of decay data of [18] for different systems in logarithmic scale.

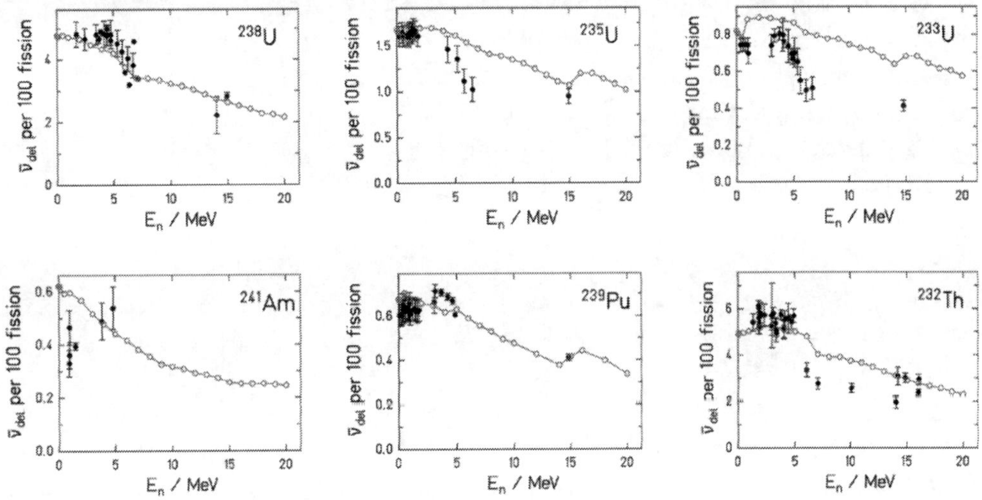

Figure 10. Energy dependence of the delayed neutron yield in neutron-induced fission of several systems. The data (black symbols) are compared with GEF results (red symbols). Figure from [6].

5 Conclusion

Manifold and complex structural features appear in light-particle production accompanying nuclear fission. They appear as local discontinuities, deviations from smooth trends or peak structures in different observables. Calculations that are able to reproduce these features and to make reasonable predictions in cases of scarce or missing experimental information must be based on a comprehensive theoretical model. In this contribution, the GEF model is given as an example and compared with experimental data for a number of selected observables.

Due to the underlying concepts and laws of general validity, and the consistent coverage of practically all fission quantities, it provides reliable results over a large range of systems (in view of nuclear composition and excitation energy), including the structural effects, with a unique description, when the parameters are determined on the basis of a limited amount of experimental data.

References

[1] D.G. Madland, A.C. Kahler, Nucl. Phys. A **957** (2017) 289
[2] I. Stetcu, P. Talou, T. Kawano, M. Jandel, Phys. Rev. C **90** (2014) 024617
[3] P. Talou, T. Kawano, I. Stetcu, Nucl. Data Sheets **118** (2014) 195
[4] O. Litaize, O. Serot, L. Berge, Eur. Phys. J. A **51** (2015) 177
[5] R. Vogt, J. Randrup, Phys. Rev. C **96** (2017) 064620
[6] K.-H. Schmidt, B. Jurado, Rep. Progr. Phys. **81** (2018) 106301
[7] N. Schunck, L.M. Robledo, Rep. Prog. Phys. **79** (2016) 116301
[8] Zi-Ao Wang, Junchen Pei, Yue Liu, Yu Qiang, Phys. Rev. Lett. **123** (2019) 122501
[9] K.-H. Schmidt, B. Jurado, C. Amouroux, C. Schmitt, Nucl. Data Sheets **131** (2016) 107
[10] U. Mosel, H.W. Schmitt, Phys. Rev. C **4** (1971) 2185
[11] B.D. Wilkins, E.P. Steinberg, R.R. Chasman, Phys. Rev. C **14** (1976) 1832
[12] C. Böckstiegel et al., Nucl. Phys. A **802** (2008) 12
[13] G. Scamps, C. Simenel, Nature **564** (2018) 382
[14] A.A. Naqvi et al., Phys. Rev. C **34** (1986) 218
[15] A.R. Junghans et al., Phys. Lett. B **670** (2008) 200
[16] J.S. Batra, Raj K. Gupta, Phys. Rev. C **43** (1991) 1725
[17] A. Gatera et al., Phys. Rev. C **95** (2017) 064609
[18] M. Estienne et al., Phys. Rev. Lett. **123** (2019) 022502

Fission in a microscopic framework: From basic science to support for applications

Ionel Stetcu[1,*], *Aurel* Bulgac[2], *Shi* Jin[2], *Kenneth J.* Roche[2,3], and *Nicolas* Schunck[4]

[1]Los Alamos National Laboratory, Los Alamos, New Mexico, 87545, USA
[2]University of Washington, Seattle, Washington 98195-1560, USA
[3]Pacific Northwest National Laboratory, Richland, WA 99352, USA
[4]Lawrence Livermore National Laboratory, Livermore, California 94551, USA

Abstract. Recent developments, both in theoretical modeling and computational power, have allowed us to make progress on a goal not fully achieved yet in nuclear theory: a microscopic theory of nuclear fission. Even if the complete microscopic description remains a computationally demanding task, the information that can be provided by current calculations can be extremely useful to guide and constrain more phenomenological approaches, which are simpler to implement. First, a microscopic model that describes the real-time dynamics of the fissioning system can justify or rule out some of the approximations. Second, the microscopic approach can be used to obtain trends, e.g., with increasing excitation energy of the fissioning system, or even to compute observables that cannot be otherwise calculated in phenomenological approaches or that can be hindered by the limitations of the method. We briefly present in this contribution the time-dependent superfluid local density approximation (TDSLDA) approach to nuclear fission, approach that has become a very successful theoretical model in many areas of many-body research. The TDSLDA incorporates the effects of the continuum, the dynamics of the pairing field, and the numerical solution is implemented with controlled approximations and negligible numerical errors. The main part of the current contribution will be dedicated to discussing the method, and recent results concerning the fission dynamics. In addition, we present results on the excitation energy sharing between the fragments, which are in agreement with a qualitative conclusions extracted from a limited number of experimental measurements of properties of prompt neutrons.

1 Overview and motivation

Fission is a complex many-body process that involves the shape evolution of a nuclear system from a compact object into two (or more) fission fragments. Phenomenological approaches to the description of fragment mass and charge distribution have reported a very good agreement with available experimental data [1–6]. In recent years, a host of developments, both in theoretical modeling and computational power, have allowed progress towards a more microscopic theory of nuclear fission [7–16].

*e-mail: stetcu@lanl.gov

An *ab initio* approach to the fission process, which starts from the inter-neutron interaction is well beyond not only short-term, but also long-term capabilities. Therefore, the method of choice for such processes is the density functional theory with different approximations for superfluid systems like Bardeen-Cooper-Schrieffer (BCS) or superfluid local density approximation (SLDA) [17, 18]. Our group has used extensively the time-dependent superfluid approximation (TDSLDA) [19], which is particularly efficient to implement on today's supercomputers that use graphic accelerators. This approach has been very successful in describing a large number of phenomena and properties such as the ground state energies of quantum systems, pairing gaps, collective modes, quantized vortices, *etc.* [16, 19–23]. In nuclear physics, in addition to fission studies, TDSLDA has been applied to studies of the excitation of collective modes in deformed open-shell nuclei [24], relativistic Coulomb excitations [25], and collisions of heavy nuclei [26].

The description of the fission process does not stop with the formation of the fission fragments and the prompt neutron and gamma emission: the daughter nuclei, which are highly excited and far from stability, continue to beta decay and emit delayed neutrons and gamma rays until they reach the stability region. Because such processes are much slower that the dynamics from saddle-to-scission, their treatment is decoupled from the formation of the fission fragments. However, the properties of the emerging fragments define the properties of prompt fission neutrons and gammas, which have been extensively measured. Thus, indirectly, these measurements can validate information about the fission fragments. The main caveat is that other approximations are involved in the description of prompt fission neutrons and gammas using phenomenological approaches [27–30].

An assumption used in all codes that compute the properties of prompt fission particles is the fact that the latter are emitted from the fully accelerated fission fragments, thus ignoring the possibility of neutron emission at scission or during acceleration. Is this a good approximation? Some phenomenological models predict that the fraction of prompt neutrons emitted at scission is significant and cite experimental evidence to support the claim [31–35]. A microscopic approach can offer support to such phenomenological models. In the microscopic TSDLDA and related approaches the fission dynamics, including any neutrons emitted at scission or during acceleration, can be observed in time.

Even if the assumption that the neutrons are emitted from the fully accelerated fragments is approximately correct, one has to keep in mind that in experiments the fragment properties cannot be measured before the emission of prompt neutrons and gamma rays. Thus, all the information regarding the excitation or spin distributions in fission fragments can only be inferred from measuring the properties of prompt neutron and gamma rays. Modelers are thus forced to introduce corrections that involve assumptions regarding neutron emission in order to extract the mass of the post-neutron emission mass distributions [36]. In this context, models that constrain quantities like the excitation energy sharing between fragments or their spin play a crucial role. And while models that do not allow for a full separation of the fission fragments have been advanced with promising results [37], an approach that allows for the computation of the excitation energies in each fragment can significantly reduce the theoretical uncertainties. We have recently established [16] that the properties of the fission fragments are truly established only when the separation between the tips of the fission fragments is of the order of $4 \ldots 5$ fm, which brings into question the validity of the phenomenological models used so far.

Even though the work and features of the fission process presented here mostly rely on the microscopic TDSLDA, the main message of this contribution is not necessarily to advocate toward one direction or the other, as both phenomenological and microscopic approaches have their advantages and disadvantages. Rather, we should envision a synergistic picture, in which the microscopic models can inform the more phenomenological ones on the validity of approximations involved, or provide additional information on trends and other quantities that could be challenging to obtain otherwise. And, more importantly, significant applications of the fission process extend from national security to energy generation, or to the origin of elements in astrophysics. Thus, the need for guidance from theoretical models is urgent, and the best approach would be to combine both phenomenological and microscopic approaches that deliver relevant information for those applications.

2 Brief theoretical background

We follow the dynamics of the nuclear system from compact configurations to separated fragments by solving in coordinate space the well known (self-consistent) time-dependent Schrödinger equation

$$i\hbar\frac{\partial}{\partial t}\Psi(\vec{r},t) = H(\rho_p,\rho_n,\cdots)\Psi(\vec{r},t), \qquad (1)$$

where $\Psi(\vec{r},t)$ are the 4-component quasiparticle wave functions, for both protons and neutrons, and the 4×4 Hamiltonian matrix operator H depends upon the instantaneous neutron and proton densities (normal, anomalous, spin, and current densities). H is derived from a given nuclear energy density functional by taking a functional derivative with respect to the density matrix and has a block structure with particle-hole and particle-particle components. For details see Refs. [16, 38]. The many-body wavefunction is approximated by a single Slater determinant composed of all the quasiparticle wavefunctions. In current simulations we include all quasiparticles that correspond to positive eigenvalues of the Hamiltonian H in the presense of constraints. During the entire evolution, the many-body solution remains a quasiparticle Slater determinant, even when the full separation of the fragments is achieved. Thus, in the spirit of Kohn and Sham, the only physical quantities that can be obtained are one-body observables, in particular densities and currents, which can be used to calculate the properties of the fission fragments.

In numerical calculations, we discretize Eq. (1) on a rectangular lattice, characterized by N_x, N_y and N_z points on directions x, y and z, respectively. In order to capture the dynamics of the system until the two fragments are fully separated, we have employed boxes with dimensions $N_x = N_y = 24$ and $N_z = 48$, with lattice constant $l = 1.25$ fm on all directions, which is equivalent with a momentum cutoff of about 500 MeV, sufficiently high to describe low-energy physics. However, calculations are in progress to reduce the lattice constant by 40%, which in turn requires performing even larger-scale calculations in order to preserve physical dimensions of the rectangular box in which the properties of the two fragments can be calculated after considerable separation.

In order to evolve the nuclear system in time, we employ the fifth order numerical method Adams-Bashford-Milne [39]. In this approach, the self-consistent Hamiltonian in Eq. (1) is applied twice per time step, with errors of order $O(\Delta t^6)$, where Δt is the numerical integration step in time. The derivatives are efficiently calculated via Fourier transforms, using GPU accelerators. On systems with GPUs, the bottleneck is often the need to transfer large amounts of data between CPUs and GPUs. We minimize the amount of data exchanged, by only transferring the densities for reduction over CPUs using MPI calls, and recompute the potentials on the fly on each GPU. This ensures almost perfect weak and strong scaling properties.

The initial states are taken beyond the outer barrier on the zero temperature potential energy surface defined by constraints on quadrupole and octupole moments, and evolved using Eq. (1) until the system splits into two fragments with their centers of mass separated by 30 fm. In general, we chose several points on the energy surface with small variances around an average value, so we simulate to some extent the compound nucleus at a chosen energy.

For example, in Ref. [16], we have used two sets of asymmetric configurations for each density functional parameterization, with a spread of 2 – 3 MeV around an average value. During the entire evolution, the solution remains one single quasiparticle Slater determinant. The properties of the two fragments are calculated by dividing the simulation box in two, and using the one-body densities in each half. Each fragment's energy can be obtained using the same functional used to evolve the full system, while their mass and charge are obtained by simply integrating the total and proton densities respectively, when the distance between the two fragments has reached the desired value. The excitation energies are simply computed by subtracting from the total energy of each half of the box the kinetic energy of each fragment and the ground state energy calculated with fractional neutron and proton numbers with the same functional. All these quantities are calculated as a function of time, and we can evaluate a lower value for uncertainties by looking at the behavior with time. Because we see variations of about a couple of MeV in the excitation energy results, we will run these calculations in boxes that allow for a larger separation. Thus, these results should be taken as preliminary. The collective flow energy is calculated by integrating the local kinetic energy

$$E_{\rm coll}(t) = \sum_{i=p,n} \int d^3\vec{r} \frac{|\vec{p}_i(\vec{r},t)|^2}{2M_n\rho_i(\vec{r},t)}, \tag{2}$$

where M_n is the nucleon mass, ρ_p, ρ_n the proton and neutron particle number densities, and p is the proton or neutron momentum, which can be expressed in terms of the current density. The collective kinetic energy is small and almost constant from saddle to close to scission, in the $1-2$ MeV range. A comprehensive study of fission properties was presented in Ref. [16], and work is in progress to address the issue of the excitation energy sharing as a function of energy.

We can use as input in simulations any Skyrme-type density functional. In the pioneering work presented in Ref. [10], we have opted for the generic SLy4 functional, that is not particularly well suited for fission calculations. In our latest work [16], we have used two functionals that provide a better description of fission barriers. However, in general we obtain similar results, with the exception of the time from saddle to scission, which can exhibit strong dependence on the choice of the density functional.

TDSLDA requires in general more extensive computational resources than other approaches to fission. However, we perform routinely such calculations, as they are well within existing computer power. Our codes can efficiently run on leadership-class machines Summit and Sierra at Oak Ridge or Lawrence Livermore National Laboratories respectively, Piz-daint (CSCS Switzerland), as well as on smaller machines like a GPU cluster at Los Alamos National Laboratory or Tsubame at Tokyo Institute of Techmology in Japan.

3 Selected results

We have recently published a series of results of the TDSLDA simulations [16]. In contrast with Ref. [10], where only a limited number of initial conditions and different pairing correlations for the same energy functional were investigated, in Ref. [16] we started with a larger number of initial conditions considering different points on the potential energy surface. The starting states were chosen to have a large spread in quadrupole deformation (Q_{20}) and mass asymmetry (Q_{30}), with similar excitation energies with respect to the ground state. After performing the TDSLDA simulation, we found that, independently of the functional used, the fission fragments emerge with similar properties (mass, charge, deformations, etc.). In all calculations, including those reported in Ref. [10], TDSLDA is consistent with expectations that the light fragment emerges deformed, while the heavy fragment is closer to spherical shape with very weak or collapsed pairing field, as it is expected to be close to a closed shell configuration. We have also shown that immediately after scission, the fragments have more elongated shapes, which quickly relax. How far after scission we continue to evolve the system, depends on the available computational resources. As noted in the previous section, we have usually evolved the system until the distance between the center of mass of the two fragments reached 30 fm. More recently we have extended the evolution to the point where the fragments are separated by 80-90 fm. While these calculations are very challenging, they are necessary as we have observed a continuous energy exchange between the two fragments. This could be due to Coulomb interactions, pairing correlations which are long range, and/or emission of neutrons, that would be free to bounce between the two halves of the simulation box, thus mimicking the energy exchange. Finally, we have recently opted to further reduce the lattice size to 1 fm from 1.25 in previous calculations, as the total energy of the system is conserved with much better accuracy [40]. All these efforts are to reduce as much numerical errors and detect subtle effects, such as neutron emission at scission or shortly after.

The relatively small effect of the initial state on the final properties of the fission fragments can be understood in terms of the strong one-body dissipation that is present in the model. As a consequence, in the dynamics from saddle-to-scission, the system behaves as a viscous fluid, with the trajectory following predominantly the steepest descent, and (collective) flow energy that remains almost constant over the duration of the motion until scission. This has two consequences: (i) invalidates adiabatic approaches in which the collective potential energy is transformed into collective kinetic energy, and (ii) provides justification to the overdamped Brownian motion approach to fission yields pioneered by Möller and Randrup [1, 2]. We have further investigated dissipation by performing a theoretical experiment, in which we apply quadrupole kicks to the nuclear system, and observe the behavior of the flow energy [16]. We have observed that due to the strong one-body dissipation, the collective kinetic energy transferred to the system is very quickly converted into internal excitation energy.

We have used the results in Ref. [16] to parameterize the excitation energy sharing between fragments as a function of the excitation energy in the fissioning system, which in turn can be expressed in a dependence of the ratio of temperatures of the incident neutron energy. In CGMF, the Los Alamos code that models the emission of prompt neutrons and gammas from the fission fragments [38], the energy sharing is parameterized by adjusting the ratio of temperatures in the two fragments, $R_T(A)$ to available experimental data on the average neutron multiplicity as a function of pre-neutron emission fragment mass [41]. However, this kind of data is scarce, and mostly available for spontaneous fission or neutron induced fission with thermal neutrons. At the moment, the same $R_T(A)$ is used for all incident energies, which assumes the same qualitative sharing of the excitation energy between the fragments at all incident energies. However, experimental evidence [42, 43] suggest that with increasing the neutron incident energy, most of the extra excitation energy is transferred to the heavy fragment. In order to use the information from TDSLDA calculations, we have assumed a factorization of the ratio of temperatures as follows:

$$R_T(A, E_n) = R_T(A, E_{\text{thermal}}) \cdot f(E_n), \tag{3}$$

where $f(E_n)$ was assumed a simple function of the neutron incident energy E_n. From TD-SLDA calculations, we have extracted the parameters of $f(E_n)$, and then applied an overall rescaling factor so that at thermal incident neutron energy $f(E_{\text{thermal}}) = 1$. We impose this renormalization so that we reproduce the available thermal data on average prompt neutron multiplicity as a function of mass. For the fit, we used about 30 SeaLL1 trajectories described in Ref. [16]. The consequence of this parameterization is shown in Fig. 1, where we show

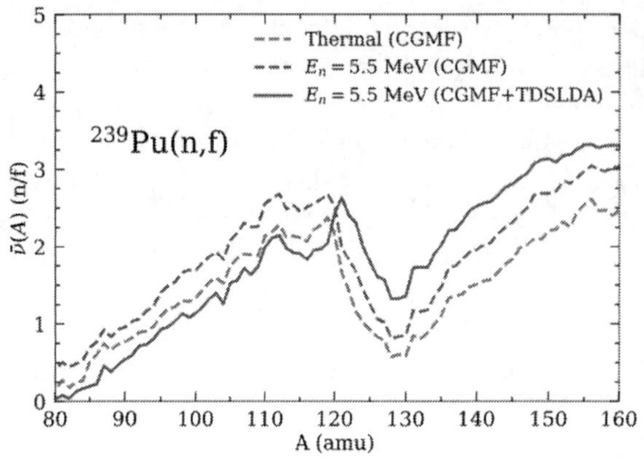

Figure 1. The average prompt neutron multiplicity as a function of fragment mass, for two different incident energies, thermal and 5.5 MeV, for the ^{239}Pu neutron-induced fission. With dashed lines we show the CGMF results in which no incident energy dependence is assumed, while with continuous line we show the 5.5 MeV CGMF results obtained using the parameterization of the incident neutron energy dependence extracted from TDSLDA calculations.

that if no energy dependence is included in CGMF calculations, increasing the incident neutron energy produces an overall increase in the neutrons emitted by both the light and the heavy fragments (the two dashed curves). However, when Eq. (3) is used, the increase in prompt average neutron multiplicity is observed mostly for the heavy fragment, consistent with available experimental data on ^{235}U and ^{237}Np neutron-induced fission [42, 43]. The small decrease in the average multiplicity for the light fragment at 5.5 MeV vs. thermal incident energy is well within conceivable experimental uncertainties, and could also be a consequence of the simplified ratio of temperature treatment employed in CGMF. Preliminary calculations using the parameterization extracted from ^{240}Pu simulations have shown a good reproduction of the measured average prompt fission neutron multiplicity as a function of mass for ^{235}U(n,f) and ^{237}Np(n,f), which suggest that this parameterization is robust.

A solid theoretical model is required to parameterize the energy sharing between the two fragments above the threshold for multi-chance fission, where neutrons can be emitted before the remaining system undergoes fission. For incident energies above second-chance fission threshold, the prompt average neutron multiplicity as a function of fragment mass is not a well defined quantity because the neutron(s) emitted before fission cannot be identified as pertaining to a fragment or the other. Hence, even if the data on average multiplicities as a function of fragment mass existed, it would be hard to disentangle the contributions from different resulting from multi-chance fission and also challenging to parameterize the energy sharing mechanism for incident neutron energies above about 5.5 MeV. TDSLDA can, in turn, provide guidance and even quantitative parameterization for this mechanism.

4 Outlook

TDSLDA has proved already an important tool in understanding the fission process. It can provide support for some of the existing approaches to fission, and invalidate others. In this contribution, the discussion was limited to a few aspects of fission dynamics, including the parameterization of the energy sharing using information from TDSLDA. However, because it can follow the dynamics of the fissioning system until full separation, TDSLDA is ideally suited to provide information on all fission fragment properties, eventually as a function of the excitation energy of the fissioning nucleus.

Current and future investigations can explore other important questions that influence the physics of prompt fission neutrons and gamma rays, which are some of the most readily available fission measurements. In particular, more work is necessary to study in more detail the neutrons that can potentially be emitted at scission, or whether part of the neutrons are emitted during acceleration. In addition, the initial spin distribution of the fission fragments can only be inferred from gamma observables, but its extraction is model-dependent [44]. TDSLDA output can be used to compute spin observables [45, 46], and an investigation is undergoing.

The need for reliable theoretical methods becomes even more clear when one considers that all the experimental measurements related to the properties of fission fragments are only possible after prompt neutron emission. Thus, primary fission fragment mass, total kinetic energy distributions or other quantities are always corrected for neutron emission, and thus is essential to understand whether neutrons are emitted at scission or during the acceleration of the fission fragments. Only models in which the dynamics of the system can be followed from before scission to well separated fragments can address such important questions.

The work of I.S. was supported by the US Department of Energy through the Los Alamos National Laboratory. Los Alamos National Laboratory is operated by Triad National Security, LLC, for the National Nuclear Security Administration of U.S. Department of Energy (Contract No. 89233218CNA000001). The work of AB and SJ was supported by U.S. Department of Energy, Office of Science, Grant No. DE-FG02-97ER41014 and in part by NNSA cooperative agreement DE-NA0003841. The TDDFT calculations have been performed by SJ at the OLCF Titan and Piz Daint and for generating initial configurations for direct input into the TDDFT code at OLCF Titan and NERSC Edison. This research used resources of the Oak Ridge Leadership Computing Facility, which is a U.S. DOE Office of Science User Facility supported under Contract No. DE- AC05-00OR22725 and of the National Energy Research Scientific computing Center, which is supported by the Office of Science of the U.S. Department of Energy under Contract No. DE-AC02-05CH11231. We acknowledge PRACE for awarding us access to resource Piz Daint based at the Swiss National Supercomputing Centre (CSCS), decision No. 2016153479. The work of KJR is supported by US DOE Office of Advanced Scientific Computing Research and was conducted at Pacific Northwest National Laboratory and University of Washington. The work of NS was supported by the Scientific Discovery through Advanced Computing (SciDAC) program funded by the U.S. Department of Energy, Office of Science, Advanced Scientific Computing Research and Nuclear Physics, and it was partly performed under the auspices of the US Department of Energy by the Lawrence Livermore National Laboratory under Contract DE-AC52-07NA27344. NS performed the calculations of the initial configurations, of the ground state and the finite-temperature properties of the FFs under the auspices of the US Department of Energy by the Lawrence Livermore National Laboratory under Contract DE-AC52-07NA27344. Some of the calculations reported here have been performed with computing support from the Lawrence Livermore National Laboratory (LLNL) Institutional Computing Grand Challenge program.

References

[1] J. Randrup, P. Möller, Phys. Rev. Lett. **106**, 132503 (2011)
[2] J. Randrup, P. Möller, A.J. Sierk, Phys. Rev. C **84**, 034613 (2011)
[3] Y. Aritomo, S. Chiba, F. Ivanyuk, Phys. Rev. C **90**, 054609 (2014)
[4] A.J. Sierk, Phys. Rev. C **96**, 034603 (2017)
[5] C. Ishizuka, M.D. Usang, F.A. Ivanyuk, J.A. Maruhn, K. Nishio, S. Chiba, Phys. Rev. C **96**, 064616 (2017)

[6] M. Verriere, N. Schunck, T. Kawano, Phys. Rev. C **100**, 024612 (2019)
[7] N. Schunck, L.M. Robledo, Rep. Prog. Phys. **79**, 116301 (2016)
[8] C. Simenel, G. Scamps, D. Lacroix, A.S. Umar, EPJ Web of Conferences **107**, 07001 (2016)
[9] W. Younes, D. Gogny, N. Schunck, *A microscopic theory of low energy fission: fragment properties*, in *Proceedings of the Fifth International Conference on ICFN5*, edited by J.H. Hamilton, A.V. Ramayya (World Scientific, 2013), p. 605
[10] A. Bulgac, P. Magierski, K.J. Roche, I. Stetcu, Phys. Rev. Lett. **116**, 122504 (2016)
[11] D. Regnier, N. Dubray, N. Schunck, M. Verriere, Phys. Rev. C **93**, 054611 (2016)
[12] H. Goutte, J.F. Berger, P. Casoli, D. Gogny, Phys. Rev. C **71**, 024316 (2005)
[13] J. Sadhukhan, W. Nazarewicz, N. Schunck, Phys. Rev. C **93**, 011304 (2016)
[14] G. Scamps, C. Simenel, Nature **564**, 382 (2018)
[15] A. Bulgac, S. Jin, I. Stetcu, Phys. Rev. C **100**, 014615 (2019)
[16] A. Bulgac, S. Jin, K.J. Roche, N. Schunck, I. Stetcu, Phys. Rev. C **100**, 034615 (2019)
[17] A. Bulgac, Phys. Rev. C **65**, 051305(R) (2002)
[18] A. Bulgac, Phys. Rev. A **76**, 040502 (2007)
[19] A. Bulgac, Ann. Rev. Nucl. Part. Sci. **63**, 97 (2013)
[20] A. Bulgac, Y.L. Luo, K.J. Roche, Phys. Rev. Lett. **108**, 150401 (2012)
[21] A. Bulgac, M.M. Forbes, M.M. Kelley, K.J. Roche, G. Wlazłowski, Phys. Rev. Lett. **112**, 025301 (2014)
[22] G. Wlazłowski, W. Quan, A. Bulgac, Phys. Rev. A **92**, 063628 (2015)
[23] G. Wlazłowski, K. Sekizawa, M. Marchwiany, P. Magierski, Phys. Rev. Lett. **120**, 253002 (2018)
[24] I. Stetcu, A. Bulgac, P. Magierski, K.J. Roche, Phys. Rev. C **84**, 051309(R) (2011)
[25] I. Stetcu, C. Bertulani, A. Bulgac, P. Magierski, K.J. Roche, Phys. Rev. Lett. **114**, 012701 (2014)
[26] P. Magierski, K. Sekizawa, G. Wlazłowski, Phys. Rev. Lett. **119**, 042501 (2017)
[27] O. Litaize, O. Serot, L. Berge, Eur. Phys. Jour. A **51**, 177 (2015)
[28] K.H. Schmidt, B. Jurado, C. Amouroux, C. Schmitt, Nuclear Data Sheets **131**, 107 (2016)
[29] J. Verbeke, J. Randrup, R. Vogt, Comp. Phys. Comm. **222**, 263 (2018)
[30] P. Talou, R. Vogt, J. Randrup, M.E. Rising, S.A. Pozzi, J. Verbeke, M.T. Andrews, S.D. Clarke, P. Jaffke, M. Jandel et al., Eur. Phys. Jour. A **54**, 9 (2018)
[31] N. Carjan, M. Rizea, Phys. Rev. C **82**, 014617 (2010)
[32] M. Rizea, N. Carjan, Nuclear Physics A **909**, 50 (2013)
[33] N. Carjan, M. Rizea, Physics Letters B **747**, 178 (2015)
[34] R. Capote, N. Carjan, S. Chiba, Phys. Rev. C **93**, 024609 (2016)
[35] N. Carjan, M. Rizea, Phys. Rev. C **99**, 034613 (2019)
[36] D.L. Duke, F. Tovesson, A.B. Laptev, S. Mosby, F.J. Hambsch, T. Bryś, M. Vidali, Phys. Rev. C **94**, 054604 (2016)
[37] M. Albertsson, B.G. Carlsson, T. Døssing, P. Möller, J. Randrup, S. Åberg, *Microscopic treatment of energy partition in fission* (2018), 1811.02283
[38] P. Talou, I. Stetcu, P. Jaffke, M. Rising, A. Lovell, T. Kawano, Computer Physics Communications **269**, 108087 (2021)
[39] R.W. Hamming, *Numerical Metchods for Scientists and Engineers* (Dover, 1986)
[40] S. Jin, K.J. Roche, I. Stetcu, I. Abdurrahman, A. Bulgac, Computer Physics Communications **269**, 108130 (2021)

[41] B. Becker, P. Talou, T. Kawano, Y. Danon, I. Stetcu, Phys. Rev. C **87**, 014617 (2013)
[42] R. Müller, A.A. Naqvi, F. Käppeler, F. Dickmann, Phys. Rev. C **29**, 885 (1984)
[43] A.A. Naqvi, F. Käppeler, F. Dickmann, R. Müller, Phys. Rev. C **34**, 218 (1986)
[44] I. Stetcu, P. Talou, T. Kawano, M. Jandel, Phys. Rev. C **88**, 044603 (2013)
[45] G.F. Bertsch, T. Kawano, L.M. Robledo, Phys. Rev. C **99**, 034603 (2019)
[46] A. Bulgac, Phys. Rev. C **100**, 034612 (2019)

Overview of two deterministic modelings for prompt emission in fission

Anabella Tudora

University of Bucharest, Faculty of Physics, 077125 Bucharest-Magurele, Romania

Abstract. Two models with a deterministic treatment of prompt emission in fission were developed at the University of Bucharest. Both models work with the same ranges of initial fragments and total kinetic energy and they use the same partition of the total excitation energy at full acceleration based on modelling at scission. The main difference between these modelings regards the prompt emission treatment itself. I.e. the Point-by-Point (PbP) model uses a global treatment of sequential emission while the other modeling is based on an event-by-event treatment of sequential emission. Both models are submitted to a rigorous validation. This paper focuses on model results of different prompt γ-ray quantities, which describe very well the existing experimental data. A new method to calculate prompt γ-ray spectra, including a global treatment based on the distribution of prompt γ-ray energy per quanta, is proposed.

1 Introduction

Two deterministic modelings of prompt emission in fission were developed at the University of Bucharest. Chronologically the first developed model was Point-by-Point (PbP), results of this model being reported during the time in many publications, see e.g. Ref. [1] which includes a comprehensive description, as well as the references therein. During the time the PbP treatment has served to obtain the average values of input parameters for prompt emission models either based on the most probable fragmentation approach (the Los Alamos (LA) model) or on the multi-modal fission concept, i.e. a most probable fragmentation associated to each fission mode. The systematic of input parameters of the LA model [2] is based on the PbP treatment, too. The PbP model was also used for nuclear data evaluations (e.g. Ref. [3] and references therein), the investigation of even-odd effects in prompt emission (e.g. Ref. [4–6] and references therein) and for the prediction of ν(A) at incident neutron energies (En) where multiple fission chances are involved [7, 8].

A second model of prompt emission was developed more recently [9], having as initial scope to obtain a general form of the residual temperature distribution P(T). The application of this model to 49 fission cases leaded to interesting systematic behaviours of different quantities characterizing the sequential prompt emission and the residual nuclei (resulting from the sequential emission) [10]. One of these systematics, i.e. the one regarding the ratios of the residual temperature (following the successive emission of each neutron) to the temperature of initial fragments (before prompt neutron emission) allowed to obtain a general form of the residual temperature distribution for each emission sequence and the inclusion of sequential emission into the LA model [10].

The main difference between these modelings regards the prompt emission itself. The PbP model uses a global treatment of sequential emission based on the residual temperature distribution on which the centre-of-mass energy spectrum of prompt neutrons at a given temperature is integrated. The other model [9, 11] includes a sequential emission treatment based on iterative equations of the residual temperature for each emission sequence associated to an initial fragment at a given value of TKE.

The similarities between these models consist of the use of the same range of initial fragmentations deterministically constructed and the same partition of the total excitation energy (TXE) based on modelling at scission. Other similarities and differences between these models, regarding the primary model results (i.e. multi-parametric matrices of different prompt emission quantities), the prescriptions used for the compound nucleus cross-sections of the inverse process of neutron evaporation from initial and residual fragments $\sigma_c(\varepsilon)$ and the level density parameters of initial and residual fragments, as well as the input model parameters are mentioned in the next section.

The results of both models, which were reported up to now, mainly refer to prompt neutrons. For this reason this work focuses on model results of different prompt γ-ray quantities and their correlation with the prompt neutron multiplicity.

2 Similarities and differences between these modelings

Both models use the same fragmentation range deterministically constructed as follows [1, 9]. The mass range of initial fragments (before prompt neutron emission) is going from symmetric fission up to a very asymmetric split (with a step of 1 mass unit). For each mass number A three or five charge numbers Z are taken as the nearest integer values above and below the most probable charge Zp(A) taken as the unchanged charge distribution $Z_{UCD}(A)$ corrected with the charge polarization $\Delta Z(A)$. For each fragment A, Z the calculations are done at TKE values covering a large range (e.g. from 100 to 200 MeV) with a step size of 2 or 5 MeV.

Both models use the same TXE partition based on modelling at scission ([1, 3] and references therein). This consists of the calculation of the extra-deformation energy of initial fragments at scission with respect to the full acceleration and the partition of the available excitation at scission (obtained by subtracting the extra-deformation energies from TXE) between complementary nascent fragments under the assumption of statistical equilibrium at scission and fragment level density in the Fermi-gas regime.

The main difference in principle regards the treatment of sequential emission. The PbP model uses a global treatment of sequential emission which is based on the residual temperature distribution P(T). The prompt neutron spectrum in the centre-of-mass frame corresponding to an initial fragment A, Z at a given TKE value is obtained by integrating the evaporation spectrum at a given temperature over P(T). The PbP computer code allows the use of different forms of P(T) either analytical expressions or numerical data (resulting from sequential emission calculations and provided as input files) [1].

The sequential emission modelling is based on recursive equations of residual temperature following the successive emission of each prompt neutron from each initial fragment A, Z of the fragmentation range at each value of the TKE range [9].

Different prescriptions for $\sigma_c(\varepsilon)$ and the level density parameter can be used in both models. In the case of PbP $\sigma_c(\varepsilon)$ are provided by optical model calculations using phenomenological optical potential parameterisations adequate for nuclei appearing as fission fragments. The use of analytical expressions or even a constant σ_c is possible, too. In the case of sequential emission treatment an analytical expression of $\sigma_c(\varepsilon)$ of initial and residual fragments (which depends on the s-wave neutron strength function provided by systematics) is currently employed (see Ref. [9] for details). The use of constant sc is also possible.

Usually energy-dependent level density parameters (provided by the super-fluid model) are used in the PbP model. In the case of sequential emission treatment, the transcendent successive equations of residual temperature can be solved only for non-energy dependent level density parameter prescriptions. In this treatment level density parameters provided by the Egidy-Bucurescu systematic for the BSFG model are used because they are close to the ones given by the super-fluid model for the majority of initial and residual fragments [9].

The primary results of the PbP model are multi-parametric matrices of different prompt emission quantities, generically labelled q(A,Z,TKE) (e.g. ν(A,Z,TKE), $<\varepsilon>$(A,Z,TKE), $\Phi(\varepsilon$,A,Z,TKE), Eγ(A,Z,TKE) etc.). The primary results of the sequential emission modelling are multi-parametric matrices of different quantities corresponding to each emission sequence k associated to an initial fragment A, Z at a given TKE value, i.e. q_k(A,Z,TKE). Multi-parametric matrices similar to the ones provided by the PbP model are obtained by averaging q_k(A,Z,TKE) over the number of sequences corresponding to an initial fragment A, Z at a given TKE value.

The secondary results of both modellings consist of single distributions of different prompt emission quantities (q(A), q(Z), q(TKE)) and total average quantities <q>. They are obtained by averaging the primary results q(A,Z,TKE) over fission fragment distributions Y(A,Z,TKE). These distributions are taken as Y(A,Z,TKE) = p(Z,A)Y(A,TKE) in which the isobaric charge distribution p(Z,A) is a Gaussian function centred on the most probable charge Zp(A) and Y(A,TKE) are experimental data.

Regarding the input parameters: to provide the primary results (i.e. the multi-parametric matrices q(A,Z,TKE)) both modellings do not use free or adjustable parameters. They need only data from recommended nuclear data libraries of RIPL1-3 (e.g. mass excesses, shell corrections, $\beta 2$ deformations, optical model parameterisations). This fact assures the possibility of prediction. For secondary results (i.e. average quantities as a function of A, of TKE etc.), experimental data of Y(A,TKE) are needed as input.

3 Model validation based on prompt γ-ray results

As it was already mentioned in Ref. [12], the main validation of a prompt emission model consists of the comparison of its primary results (i.e. multi-parametric matrices) with existing experimental data. This is a validation of the model itself because the Y(A,TKE) distribution is not involved.

The comparison of secondary model results (i.e. single distributions of different prompt emission quantities and total average quantities) with experimental data which are available in a much larger amount than in the case of multi-parametric data, is considered as a secondary validation of the prompt emission model together with the Y(A,TKE) distribution used [12].

The primary and secondary validations of both models reported up to now, e.g. in Refs. [1, 9, 11, 12], were mainly based on recent experimental data referring to prompt neutrons (e.g. the recent ν(A,TKE) data measured by Göök *et al.* for ^{252}Cf(SF) [13] and ^{235}U(n,f) [14], the centre-of-mass energy spectrum $\Phi(\varepsilon)$ for the light and heavy fragment groups of ^{235}U(n,f) [14] etc.), which were very well described by both model results.

For this reason the present paper focuses on the validation of both models based this time on prompt γ-ray results (obviously obtained concomitantly with the prompt neutron results, in the same run), which are compared with previous and recent prompt γ-ray data.

An example of primary validation is illustrated in Fig. 1 where the experimental prompt γ-ray energy matrix Eγ(A,TKE) of ^{235}U(n$_{th}$,f) [15] (black symbols), in the 2D representations as Eγ as a function of A for a given TKE value (upper part) and as Eγ as a function of TKE

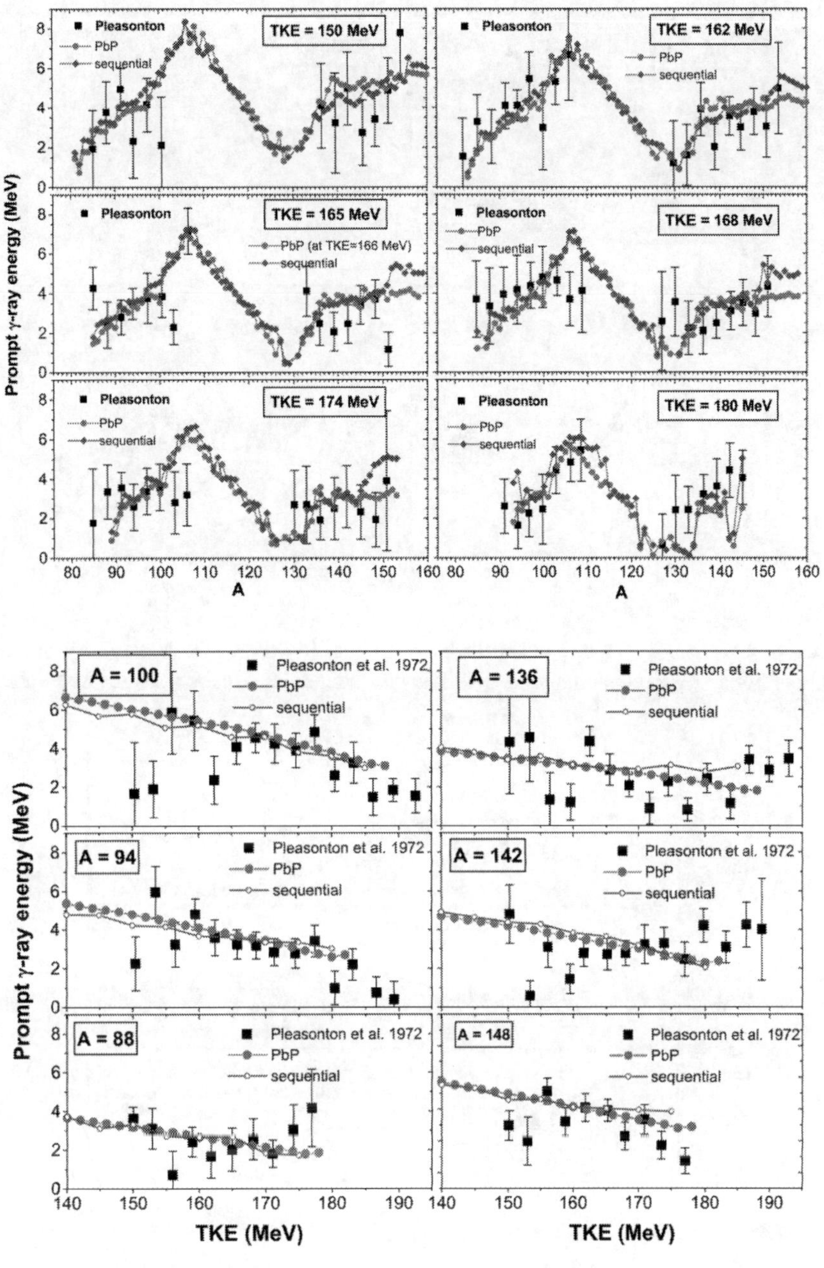

Figure 1. Eγ(A,TKE) results of PbP (red symbols) and sequential emission (blue symbols) calculations in comparison with the data of Pleasonton *et al.* [15] (black symbols) in the 2D representations as Eγ(A) at a given TKE value (upper part) and as Eγ(TKE) for a given fragment mass (lower part).

for a given fragment mass (lower part), is well described by the Eγ(A,TKE) results of both models (red symbols for PbP and blue symbols for the sequential emission modelling).

Examples of secondary validation are given in Fig. 2, where the average prompt γ-ray energy and multiplicity as a function of A (left part) and as a function of TKE (right part) for $^{235}U(n_{th},f)$ provided by the PbP (red symbols) and sequential emission (blue symbols) calculations are in agreement with the experimental data (black and grey symbols). Another

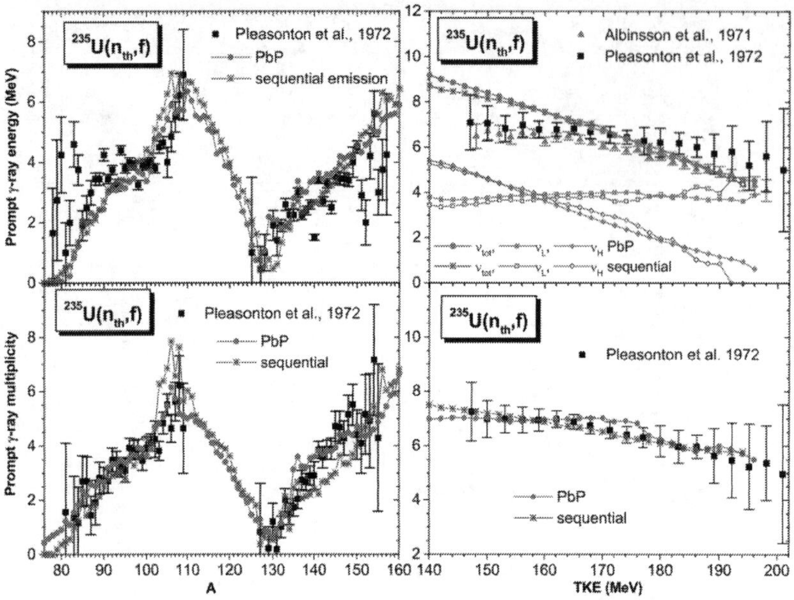

Figure 2. Eγ(A), Nγ(A), Eγ(TKE) and Nγ(TKE) results of PbP (red symbols) and sequential emission (blue symbols) for ^{235}U(n_{th},f) in comparison with experimental data (black and grey symbols).

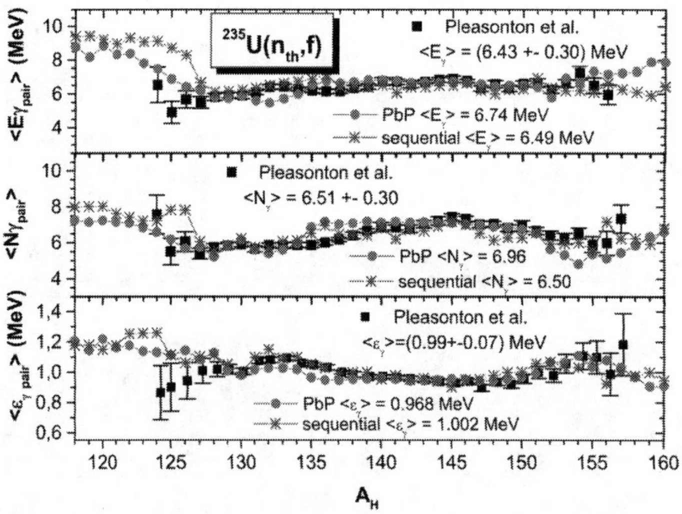

Figure 3. Experimental data of prompt γ-ray energy, multiplicity and energy per quanta of fragment pair for ^{235}U(n_{th},f) [15] (black symbols) and both model results (red and blue symbols, respectively).

example is given in Fig. 3 where the experimental data of the average prompt γ-ray energy, multiplicity and prompt γ-ray energy per quanta of the mass pair [15] are well described by the both model results, PbP and sequential emission.

Correlations between different prompt γ-ray quantities and the prompt neutron multiplicity were also investigated. The most known is the linear correlation between the prompt γ-ray energy and the prompt neutron multiplicity (which was reported during the time by many authors including us). For a nucleus fissioning spontaneously or at a given excitation energy,

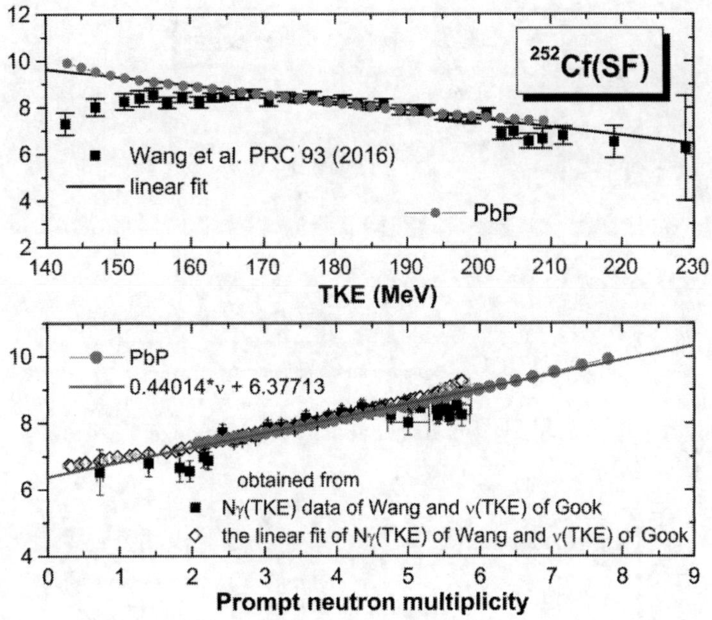

Figure 4. <Nγ>(TKE) (upper part) and the correlation <Nγ>(ν) (lower part) for ^{252}Cf(SF): the experimental data (full and open black symbols) and the PbP results (red circles).

this correlation is an obvious consequence of the linear decreasing behaviours of the average prompt neutron multiplicity and prompt γ-ray energy as a function of TKE.

The almost linear decrease of the average prompt γ-ray multiplicity as a function of TKE (exemplified for ^{235}U(n$_{th}$,f) in the right lower part of Fig. 2 and for ^{252}Cf(SF) in the upper part of Fig. 4) together with the well known linear decrease of the average prompt neutron multiplicity as a function of TKE [1, 3, 9, 11–14] leads to a linear correlation between the prompt γ-ray multiplicity and the prompt neutron multiplicity, as it can be seen in the lower part of Fig. 4 for ^{252}Cf(SF) taken as an example.

The experimental data plotted in the lower part of Fig. 4 (with full and open black symbols) were obtained from the experimental Nγ(TKE) data of Wang *et al.* [16] and their linear fit respectively, and the experimental <ν>(TKE) data of Göök *et al.* [13]. As it can be seen, these data exhibit an almost linear behaviour well described by the PbP result (red circles).

Another correlation, illustrated in Fig. 5 for the case of ^{252}Cf(SF), regards the linear behaviour of $\sigma_\gamma{}^2$ = <Eγ^2> - <Eγ^2> as a function of prompt neutron multiplicity.

The experimental data plotted with black squares in Fig. 5 were obtained from the experimental data Nγ(TKE) of Wang *et al.* [16], Eγ(TKE) of Nifenecker *et al.* [17] and ν(TKE) of Göök *et al.* [13] using the relation $\sigma_\gamma{}^2$ = 2<Eγ>^{2}/N<γ> (based on the assumption of a Poisson distribution of Nγ). As it can be seen these data are well described by the PbP result (red circles) which exhibits an obvious linear behaviour.

As an alternative to the statistical Hauser-Feshbach calculation of prompt γ-ray emission in competition with the neutron emission from many nuclei appearing as fission fragments (a great part of these nuclei having scarce or unknown level schemes) requiring a long computing time, a more simple method to calculate the prompt γ-ray spectrum (adequate for evaluation purposes) is proposed. This method consists of a global treatment following an idea which is similar to the one based on P(T) which is used for the prompt neutron spectrum. This method is based on the distribution of prompt γ-ray energy per quanta D($\varepsilon\gamma$) of the light

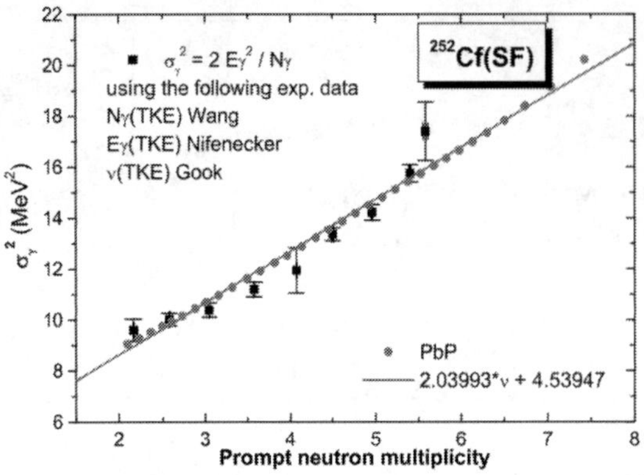

Figure 5. $\sigma_\gamma{}^2$ as a function of prompt neutron multiplicity for the case of ^{252}Cf(SF).

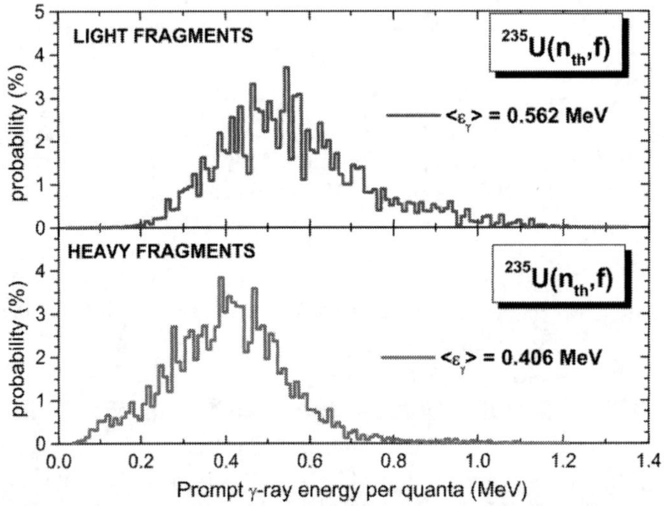

Figure 6. Distributions of prompt γ-ray energy per quanta corresponding to the light and heavy fragment groups of ^{235}U(n$_{th}$,f) resulting from the PbP treatment.

and heavy fragments on which the prompt γ-ray spectrum at a given prompt γ-ray energy per quanta $\varphi(\varepsilon\gamma,E\gamma)$ is integrated, the result being multiplied with the average prompt γ-ray multiplicity of the light and heavy fragment groups, i.e.:

$$\Phi(E_\gamma) = \overline{N}_{\gamma L} \int_0^{\varepsilon\gamma_{\max L}} D_L(\varepsilon_\gamma)\,\varphi(\varepsilon_\gamma, E_\gamma)\,d\varepsilon_\gamma \;+\; \overline{N}_{\gamma H} \int_0^{\varepsilon\gamma_{\max H}} D_H(\varepsilon_\gamma)\,\varphi(\varepsilon_\gamma, E_\gamma)\,d\varepsilon_\gamma \qquad (1)$$

Distributions of prompt γ-ray energy per quanta entering Eq. (1) are exemplified in Fig. 6 for the case of ^{235}U(n$_{th}$,f). They were obtained from the primary PbP result of $\varepsilon\gamma$(A,Z,TKE) and the fragment distribution of Ref. [18].

Figure 7 shows preliminary results of prompt γ–ray spectrum based on Eq. (1) (red lines) exemplified for ^{235}U(n$_{th}$,f) (upper part) and ^{239}Pu(n$_{th}$,f) (lower part). As it can be seen they give an overall good description of the experimental data for ^{235}U(n$_{th}$,f) measured by Oberstedt *et al.* [19] (i.e. five data sets taken from EXFOR [20], with the entries specified in

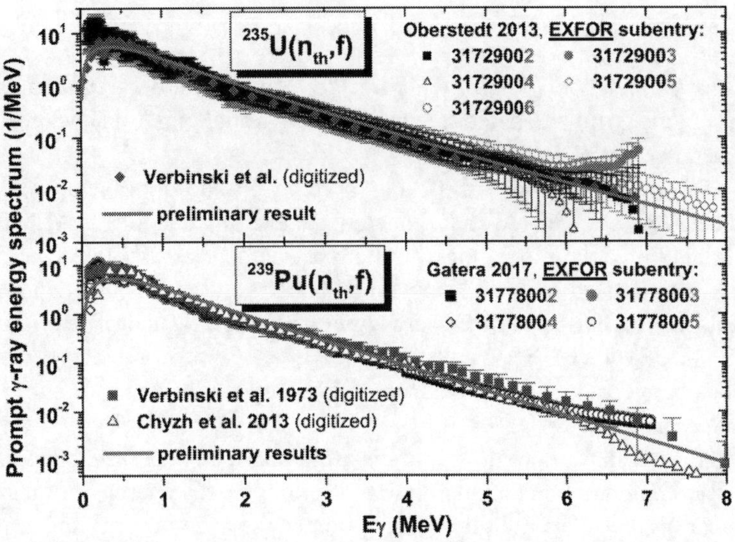

Figure 7. Preliminary prompt γ-ray spectrum results (red lines) for ^{235}U(n$_{th}$,f) (upper part) and ^{239}Pu(n$_{th}$,f) (lower part) in comparison with the experimental data [19–24] (different full and open symbols).

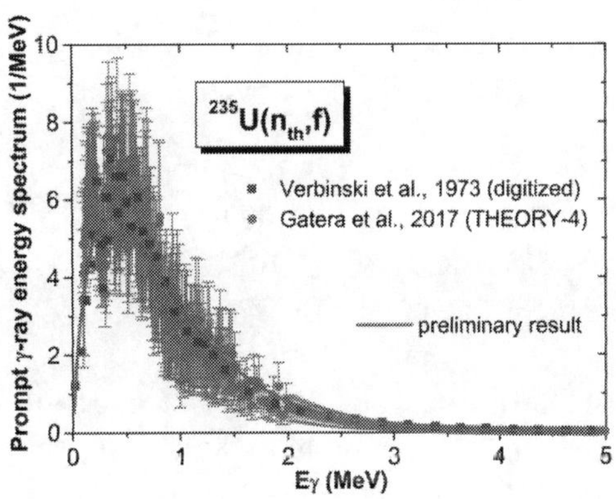

Figure 8. Focus of the low energy part of the prompt γ-ray spectrum of ^{235}U(n$_{th}$,f).

the figure legend, which correspond to the measurements using five different detectors) and Verbinski *et al.* [21] and for ^{239}Pu(n$_{th}$,f) measured by Gatera *et al.* [22] taken from EXFOR [20] (also with the entries of data sets given in the figure legend), Verbinsky *et al.* [21] and Chyzh *et al.* [23].

A focus of the low energy part of the prompt γ-ray spectrum is illustrated in Fig. 8 where the calculated spectrum of ^{235}U(n$_{th}$,f) is compared with recent experimental data [24] (grey symbols). A slight underestimation of a part of experimental data, but remaining inside the error bars, is observed at low prompt γ-ray energies below 1 MeV.

4 Conclusions

Both deterministic models of prompt emission (PbP with a global treatment of sequential emission and a model with a detailed treatment of sequential emission) were submitted to a rigorous and detailed validation including:

i) The validation of the model itself: i.e. a very good description of multi-parametric experimental data (e.g. recent ν(A,TKE) data and the existing data of Eγ(A,TKE)).

ii) The validation of the model together with a fragment distribution Y(A,TKE): i.e. a very good description of experimental single distributions and total average quantities related to prompt neutrons (e.g. ν(A), ν(TKE), $<\varepsilon>$(A), $<\varepsilon>$(TKE), P(ν), $\Phi(\varepsilon)$ etc.) and to prompt γ-rays (e.g. Eγ(A), Eγ(TKE), Nγ(A), Nγ(TKE), $<E\gamma>$, $<N\gamma>$, $< \varepsilon\gamma >$, $\Phi(E\gamma)$ etc.)

Correlations between the prompt neutron multiplicity and different prompt γ-ray quantities were emphasized, too.

The preliminary results of prompt γ-ray spectrum provided by a simple modelling including a global treatment based on the distribution of prompt γ-ray energy per quanta (provided by the PbP model) give an overall good description of the prompt γ-ray spectrum data measured for ^{235}U(n$_{th}$,f), ^{239}Pu(n$_{th}$,f) and ^{252}Cf(SF).

The determination of a general analytical form for the distribution of prompt γ-ray energy per quanta (as in the case of the triangular residual temperature distribution P(T) used in the prompt neutron spectrum calculation) is in progress.

This work was done in the frame of the Romanian Project PN-III-P4_PCE-2016-0014 (contract No. 7/2017).

References

[1] A. Tudora, F.-J. Hambsch, Eur. Phys. J. A **53**, 159 (2017)
[2] A. Tudora, Ann. Nucl. Energy **36**, 72-84 (2009)
[3] R. Capote *et al.*, Nucl. Data Sheets **131**, 1-106 (2016)
[4] A. Tudora, F.-J. Hambsch, G. Giubega, Eur. Phys. J. A **52**, 182 (2016)
[5] A.Tudora, F.-J. Hambsch, G. Giubega, I. Visan, Rom. Rep. Phys. **68**, 571-581 (2016)
[6] G. Giubega, I. Visan, A. Tudora, Rom. Rep. Phys. **68** 1024-1034 (2016)
[7] A. Tudora, F.-J. Hambsch, V. Tobosaru, Phys. Rev. C **94**, 044601 (2016)
[8] A. Tudora, F.-J. Hambsch, V. Tobosaru, Nuc. Sci. Eng. **192**, 52-69 (2018)
[9] A. Tudora, F.-J. Hambsch, V. Tobosaru, Eur. Phys. J. A **54**, 87 (2018)
[10] A. Tudora, "Sistemtic behaviours of different quantities related to sequential prompt emission in fission", ND-2019 Int. Conf. on Nuclear Data for Science and technology, May 2019, Beijing, China, proceeding in press
[11] A. Tudora, A. Matei, Rom. J. Phys. **64** (1-2) 301 (2019)
[12] A. Tudora, Eur. Phys. J. A **55**, 98 (2019)
[13] A. Göök, F.-J. Hambsch, M. Vidali, Phys. Rev. C **90**, 064611 (2014)
[14] A. Göök, F.-J. Hambsch, S. Oberstedt, M. Vidali, Phys. Rev. C **98**, 044615 (2018)
[15] F. Pleasonton, R.L. Ferguson, H.W. Schmitt, Phys. Rev. C **6**, 1023-1039 (1972)
[16] T. Wang *et al.*, Phys. Rev. C **93**, 014606 (2016)
[17] H. Nifenecker, C. Signarbieux, R. Rabinet, J. Poitou, "Neutron and gamma emission in fission" (review paper), IAEA-SM-174/207, 117-178 (1973)
[18] A. Al-Adili, F.-J. Hambsch, S. Pomp, S. Oberstedt, Phys. Rev. C **86**, 054601 (2012)
[19] A. Oberstedt *et al.*, Phys. Rev. C **87**, 051602 (2013)

[20] EXFOR Experimental Nuclear Data Library (available online at https://www-nds.iaea.org) nucleus U-235, reaction (n,f), entries 31729002-31729006 (Oberstedt *et al.*); nucleus Pu-239, reaction (n,f), entries 31778002-31778005 (Oberstedt *et al.*)
[21] V. V. Verbinski, H. Weber, R. E. Sund, Phys. Rev. C **7**, 1173-1185 (1973)
[22] A. Gatera *et al.*, Phys. Rev. C **95**, 064609 (2017)
[23] A. Chyzh *et al.*, Phys. Rev. C **87**, 034620 (2013)
[24] A. Gatera *et al.*, EPJ Web of Conferences **169**, 00003 (2018)

Possible ternary fission of super-heavy nuclei

Takahiro Wada[1,*], *Kazuki* Okada[1], *Nicolae* Carjan[2,3], and *Fedir* Ivanyuk[4]

[1]Department of Pure and Applied Physics, Kansai University, 564-8680 Suita, Osaka, Japan
[2]National Institute of Physics and Nuclear Engineering, 76900 Bucharest-Magurele, Romania
[3]Joint Institute for Nuclear Research, FLNP, 141980 Dubna, Moscow Region, Russia
[4]Institute for Nuclear Research, 03680 Kiev, Ukraine

Abstract. In super-heavy nuclei, we may expect to discover new phenomena because of the strong Coulomb force. One example is the true ternary fission. We study the sequential two binary fission of $^{300}120$ and ^{252}Cf that produces three fragments. We use the Metropolis method to estimate the probability of the second fission. The probability of the second fission is 10^{-4}–10^{-3} for super-heavy nuclei while it is 10^{-7}–10^{-6} for heavy actinide nuclei. The most probable mass division is almost symmetric in the case of $^{300}120$. We also demonstrate the applicability of the Metropolis method to a non-equilibrium process by comparing it with the Langevin equation.

1 Introduction

In super-heavy nuclei, because of the large atomic number, one may expect to discover new phenomena that we could not observe in lighter nuclei. One example is a new fission such as the true ternary fission where a massive nucleus decays into three fragments of not very different masses [1]. For actinide nuclei, ternary fission means the light-charged-particle-accompanied binary fission. It is characterized by the angular distribution of the light charged particle that is peaked around the direction perpendicular to the fission axis. It is interpreted that the light particle is emitted from the neck region so that it is accelerated by the Coulomb fields of the two heavy fragments. The possibility that a heavy excited nucleus might split into three fragments of comparable mass was considered soon after the discovery of fission. Such decays of low excited heavy nuclei have not been unambiguously observed. For thermal neutron fission of U and Pu, a relative frequency of ternary to binary events is reported to be the order of one in 10^6 [1]. There have been attempts to predict the possible combination of the three fragments for super-heavy nuclei [2, 3]. Zagrebaev *et al.* [2] found that ternary decays with the formation of two heavy and a lighter third fragment are possible for super-heavy nuclei. By assuming symmetric ternary fission, they obtained ^{132}Sn-^{32}S-^{132}Sn colinear configuration for $^{296}116$. Later, Vijayaraghavan *et al.* [3] calculated the yield of "colinear cluster tripartition" of ^{252}Cf (measured by Pyatkov *et al.* [4]) within the framework of three cluster model. They obtained ^{88}Se+^{82}Ge+^{82}Ge as one probable fragment combination. These works are essentially based on the argument of the final Q-value of the reaction. In order to obtain more quantitative estimates of ternary fission in very heavy and super-heavy nuclei, we need a dynamical approach that takes account of the fission saddle heights.

*e-mail: wadataka@kansai-u.ac.jp

It is known, especially for spontaneous fission, that shell correction energy plays an important role in determining the mass distribution of the fragments. Recently, Carjan *et al.* [5] calculated the pre-scission configuration for the binary fission of super-heavy nuclei using the microscopic-macroscopic approach. They suggested that the $N = 82$ neutron shell plays a dominant role in determining the most probable mass separation. The light fragment has the mass number around 136 and has almost spherical shape. On the other hand, the heavy fragment is found to be extremely elongated. It is possible that the elongated heavy fragment undergoes a subsequent binary fission thus producing three fragments. In this paper, we will estimate the probability of ternary fission in comparison with that of the binary fission. In Sect. 2, we will describe our framework and in Sect. 3, the results are given with discussions. In Sect. 4, we will give the summary.

2 Framework

First, we select the most probable binary mass separation and determine the nuclear shape just before scission by the pre-scission model [6]. The shape of the mother nucleus is expressed by the generalized Cassinian oval parameterization [7]. In the original Cassinian oval parameterization, the parameter α specifies the elongation of the nucleus; $\alpha = 0$ corresponds to the spherical shape and $\alpha = 1$ corresponds to the zero-neck shape [7]. In the generalized Cassinian oval parameterization, nuclear shapes are described in terms of Cassinian ovals generalized by the inclusion of additional shape parameters: α_1, α_3, α_4 and α_6. The pre-scission shape is calculated by minimizing the deformation energy with respect to ($\alpha_1, \alpha_3, \alpha_4, \alpha_6$) for $\alpha_1 = 0.98$. The deformation energy is calculated with the microscopic-macroscopic approach [8]. We show, in Fig. 1, the pre-scission shape of a super-heavy nucleus $^{300}120$. As can be seen, the light fragment of mass 136 is almost spherical while the heavy fragment is strongly deformed. Since we are interested in the probability of the second fission of the deformed fragment, the shape of the deformed fragment is fitted by the generalized Cassinian oval parameterization as shown by the red curve in Fig. 1.

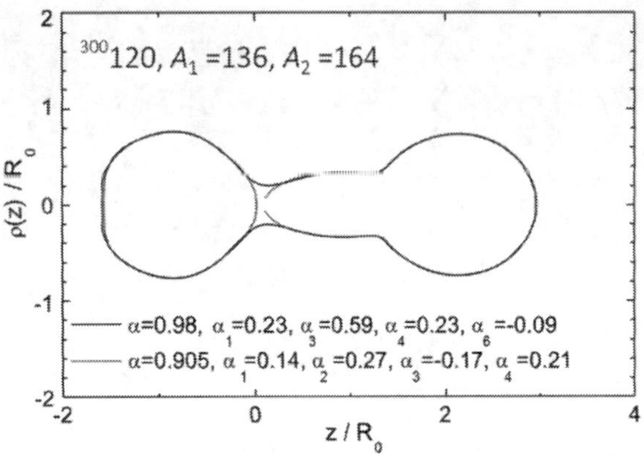

Figure 1. Pre-scission shape of the mother nucleus $^{300}120$.

The development of the shape of the deformed fragment is calculated by means of the Metropolis method. The specific algorithm introduced by Metropolis [9] has two parts: 1. Each configuration χ has a specified set of "neighbours" $\{\chi'\}$. 2. When the system is at the configuration χ, the next configuration in the sequence is sampled as follows: first, a tentative next configuration, χ', is selected randomly from the set of neighbours and it is then decided whether the system makes a "step" to that new configuration χ' or remains at the current configuration χ. The Metropolis step probability is given by

$$P(\text{step}) = \begin{cases} 1 & (W(\chi') > W(\chi)) \\ W(\chi')/W(\chi) & (W(\chi') < W(\chi)). \end{cases} \tag{1}$$

Namely, if the statistical weight of the tentative configuration is greater than that of the current configuration then the step probability is one, and in the opposite case, then the probability of taking the tentative configuration is W'/W and the probability of staying in the current configuration is 1 - W'/W. In the present work, we assume that the statistical weight is proportional to $\exp(-V(\chi)/T)$, where T is the temperature of the system, then the step probability is given as

$$P(\text{step}) = \begin{cases} 1 & (V(\chi') < V(\chi)) \\ \exp\left(-\frac{V(\chi')-V(\chi)}{T}\right) & (V(\chi') > V(\chi)). \end{cases} \tag{2}$$

Now, for the calculation of the second fission, we need the potential energy surface of the deformed fragment in the two-dimensional configuration space (α, A_L), where A_L denotes the mass number of the (future) light fragment in the second fission. We start from an initial configuration and let the samples move according to the Metropolis method. For the current position (α, A_L), we take the neighbours as (α, $A_L \pm \Delta$A) and ($\alpha \pm \Delta\alpha$, A_L). In the actual calculation, we take $\Delta A = 0.5$ and $\Delta\alpha = 0.0025$. For simplicity, we use the fixed temperature T to calculate the step probability. We continue the random walk until the sample reaches either $\alpha = \alpha_{\text{MAX}} = 1.0$ or $\alpha = \alpha_{\text{MIN}} = 0.85$. Fission probability is calculated by dividing the number of samples that reach α_{MAX} by that of all samples. The number of samples ranges from 10^7 to several 10^8 depending on the fission probability.

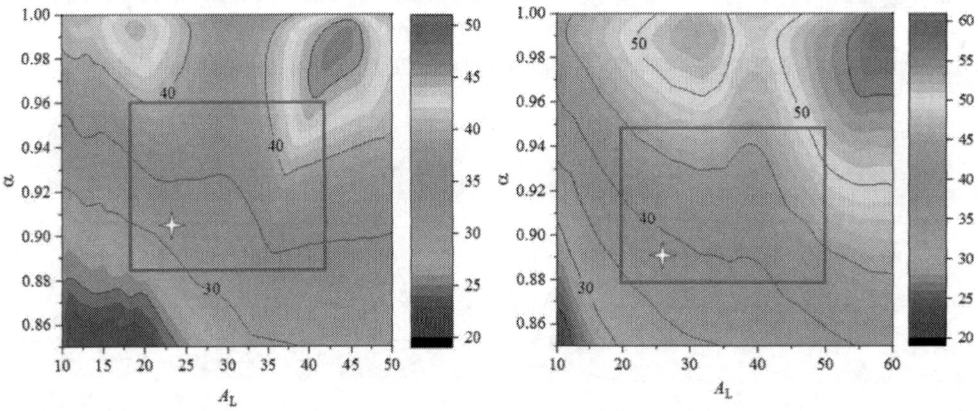

Figure 2. Potential energy surface of the deformed fragment: left panel is for Tb and right panel is for Pd.

3 Results and discussions

In this work, we calculate the probability of sequential binary fission with respect to the first binary fission to estimate the probability of observing "ternary" fission in which we observe three fragments of significant mass. We choose two systems: a super-heavy nucleus $^{300}120$ and a heavy actinide nucleus ^{252}Cf. According to the pre-scission model, the most probable mass separation is $A_1 = 136$ and $A_2 = 164$ (Tb, $Z = 65$) for the fission of $^{300}120$ and $A_1 = 132$ and $A_2 = 120$ (Pd, $Z = 46$) for ^{252}Cf. In both cases, one fragment is found to be almost spherical while the other fragment is fairly deformed as shown in Fig. 1. In order to estimate the probability of the second fission with the Metropolis method, we calculate the energy surface of the deformed fragment (^{164}Tb or ^{120}Pd) by minimizing the energy with respect to α_3 and α_4 for each set of (α, A_L). Figure 2 shows the contour plot of the deformation energy in the two-dimensional configuration space (α, A_L) for ^{164}Tb (left panel) and ^{120}Pd (right panel). The point marked with a star corresponds to the most probable initial configuration which is obtained by refitting the shape obtained with the pre-scission model. The red rectangle denotes the area from where we take the points to start the Metropolis random walk.

3.1 Probability of the second fission with respect to the first binary fission

In this subsection, we will discuss the probability of the second fission. Table 1 shows the probability of the second fission with respect to the first binary fission for nine initial configurations around the most probable configuration for the case of a super-heavy nucleus ^{300}120. The most probable initial configuration is $\alpha = 0.905$ and $A_L = 24$, as is shown by a star symbol in the left panel of Fig. 2. In the case of $T = 1.0$ MeV, the fission probability is found to be in the order of 10^{-4} and it is in the order of 10^{-3} for $T = 1.5$ MeV. Table 2 is for the case of ^{252}Cf. The most probable initial configuration is $\alpha_1 = 0.89$ and AL = 26 in this case. The probability of the second fission of ^{252}Cf is found to be in the order of 10^{-7} in the case of $T = 1.0$ MeV and in the order of 10^{-6} in the case of $T = 1.5$ MeV. In all cases, the fission probability depends significantly on the initial position. Nevertheless, we can conclude that the fission probability will be larger by a few orders of magnitude in the case of super-heavy nuclei than in the case of actinide nuclei (Cf).

Table 1. Fission probability of the fragment ^{164}Tb calculated for several initial configurations.

^{164}Tb	$T = 1.0$ MeV			$T = 1.5$ MeV		
	$A_L = 22$	$A_L = 24$	$A_L = 26$	$A_L = 22$	$A_L = 24$	$A_L = 26$
$\alpha = 0.910$	1.2×10^{-4}	4.8×10^{-4}	9.4×10^{-4}	1.3×10^{-3}	3.2×10^{-3}	5.4×10^{-3}
$\alpha = 0.905$	5.5×10^{-5}	2.3×10^{-4}	5.1×10^{-4}	7.5×10^{-4}	1.9×10^{-3}	3.4×10^{-3}
$\alpha = 0.900$	2.4×10^{-5}	1.0×10^{-4}	2.8×10^{-4}	4.2×10^{-4}	1.1×10^{-3}	2.2×10^{-3}

Table 2. Fission probability of the fragment ^{120}Pd calculated for several initial configurations.

^{120}Pd	$T = 1.0$ MeV			$T = 1.5$ MeV		
	$A_L = 24$	$A_L = 26$	$A_L = 28$	$A_L = 24$	$A_L = 26$	$A_L = 28$
$\alpha = 0.895$	6.0×10^{-7}	9.9×10^{-7}	2.2×10^{-6}	5.4×10^{-6}	8.0×10^{-6}	1.2×10^{-5}
$\alpha = 0.890$	2.8×10^{-7}	6.2×10^{-7}	1.1×10^{-6}	3.2×10^{-6}	4.9×10^{-6}	7.6×10^{-6}
$\alpha = 0.885$	1.6×10^{-7}	2.9×10^{-7}	5.9×10^{-7}	1.9×10^{-6}	3.1×10^{-6}	4.7×10^{-6}

In order to understand what determines the probability of the second fission, we perform the calculation starting from many points in (α, A_L) space. The calculation area is marked by a red rectangular in Fig. 2. The results are plotted as contour maps in Figs. 3 and 4. We plot the common logarithm of the probability, namely $\log_{10}(P_f(\alpha, A_L))$ where $P_f(\alpha, A_L)$ is the probability of the second fission with the starting point being (α, A_L). Figure 3 is for ^{164}Tb and Fig. 4 is for ^{120}Pd. The left panels show the results for $T = 1.0$ MeV and the right panels show those for $T = 1.5$ MeV. Obviously, the fission probability is larger for $T = 1.5$ MeV than that is for $T = 1.0$ MeV. On the other hand, as is seen from the figures, the structure of the contour plot is the same for $T = 1.0$ MeV and for $T = 1.5$ MeV. The temperature dependence comes from the statistical weight $\exp(-\Delta V/T)$ at each step of the Metropolis method. The feature of the plot can be understood with the two factors: the statistical weight $\exp(-(V_{\text{sad}} - V(\alpha, A_L))/T)$ and the distance between the initial and the saddle point. Here, V_{sad} denotes the potential energy at the saddle point. The saddle point lies around $(\alpha, A_L) = (0.98, 30)$ for ^{164}Tb and $(0.99, 40)$ for ^{120}Pd. The fission probability depends on the distance, because it is a non-equilibrium diffusion process.

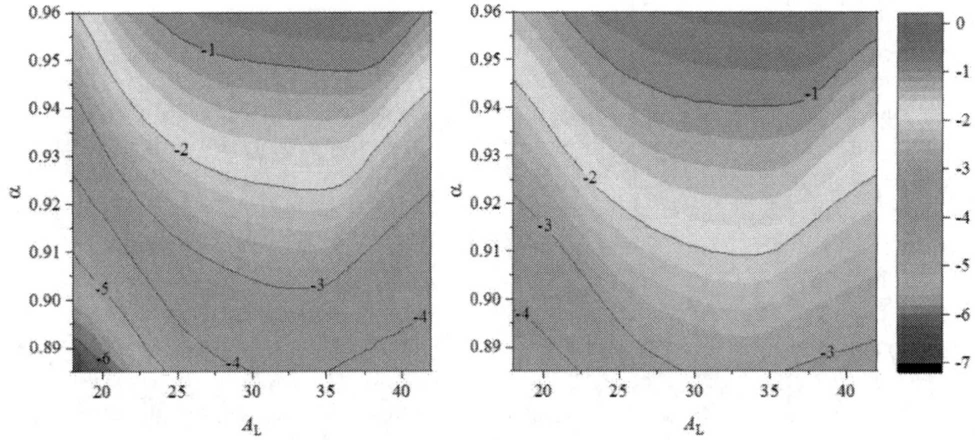

Figure 3. Fission probability $\log_{10} P_f$ of the deformed fragment ^{164}Tb as a function of the initial configuration (α, A_L). Left panel is for T = 1.0 MeV and right panel is for T = 1.5 MeV.

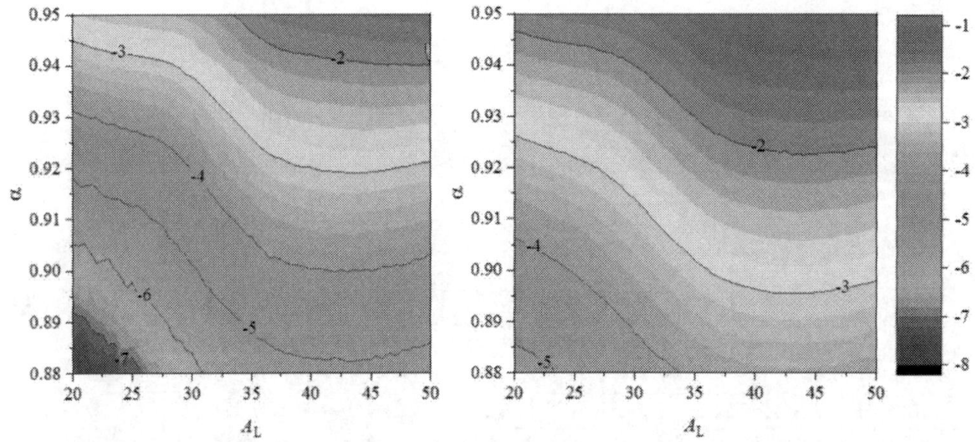

Figure 4. Fission probability $\log_{10} P_f$ of the deformed fragment ^{120}Pd as a function of the initial configuration (α, A_L). Left panel is for T = 1.0 MeV and right panel is for T = 1.5 MeV.

When we start the calculation from the points close to the most probable initial configuration, the mass of the light fragment of the second fission (at α = 1.0) is found to be close to that at the saddle point. Thus, we predict the most probable mass divisions in the ternary fission to be 136+30+134 for ^{300}120 and 132+40+80 for ^{252}Cf. Hence, in the case of superheavy nuclei, the heavy outer fragments have almost equal masses.

3.2 The Metropolis method and the Langevin approach

The Metropolis method was introduced as a means for exploring available configurations according the associated statistical weight. In order to demonstrate the applicability of the Metropolis method in calculating a non-equilibrium process, we solve a simple one-dimensional over barrier process with three methods: the Metropolis method, the reduced Langevin equation in q-space, and the Langevin equation in (p, q)-space. The Langevin equation in (p, q)-space is written as

$$\begin{aligned} \frac{dp}{dt} &= -\frac{dV}{dq} - \frac{\gamma}{m}p + R(t)\,, \\ \frac{dq}{dt} &= \frac{p}{m}\,. \end{aligned} \tag{3}$$

And the reduced Langevin equation in q-space is written as

$$\frac{dq}{dt} = \frac{1}{\gamma}\left(-\frac{dV}{dq} + R(t)\right). \tag{4}$$

Here m denotes the inertia mass, γ the friction coefficient, and V the potential. $R(t)$ denotes the random force which has the following properties;

$$\begin{aligned} \langle R(t)\rangle &= 0\,, \\ \langle R(t)R(t')\rangle &= 2\gamma T\delta(t-t')\,, \end{aligned} \tag{5}$$

where $\langle\ \rangle$ denotes the ensemble average. We assume a constant mass and friction. The potential energy is given as a connection of parabola, inverse parabola, and parabola shape which is drawn in a small panel in Fig. 5. We put a sample particle at $x = 0.5$ and calculate the development using the three methods. In the case of the Langevin equation in (p, q)-space, we consider two cases: zero initial momentum case (p=0) and thermal initial momentum distribution case (p=thermal). We set the mass $m = 1$ and the friction coefficient $\gamma = 5$. We assume a fixed temperature of $T = 1$ in this calculation.

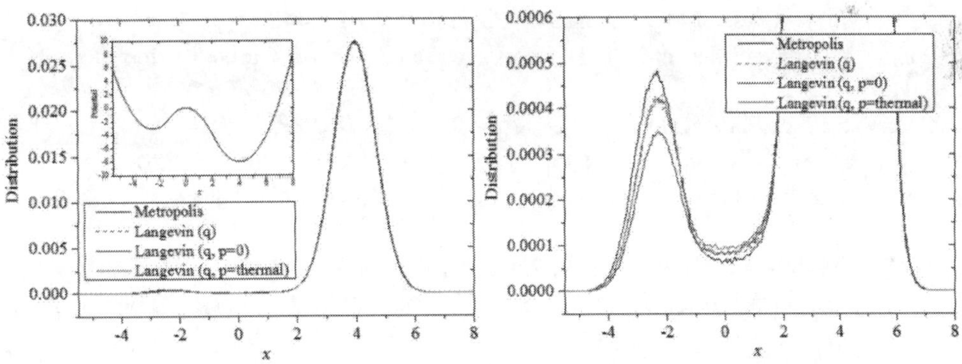

Figure 5. Comparison of the three methods to calculate the diffusion process over a barrier. Black line denotes the result with the Metropolis method, red dashed line that of the reduced Langevin equation, blue line that of the Langevin equation with zero initial momentum, and green line denotes that of the Langevin equation with thermal initial momentum distribution.

In Fig. 5, we plot the distribution at the moment when we observe the largest probability in the left potential pocket. In the left panel, we show the probability distribution in coordinate space and in the right panel, we magnify it so that we can see the difference in the left pocket region. It should be noted that if we continue the calculation for a long time, all three methods give the same distribution and the thermal equilibrium is accomplished. One can see in the left panel of Fig. 5 that the big peak in the right potential pocket is almost the same in all cases. On the other hand, in the left potential pocket, as you can see in the right panel of Fig. 5, the height of the peak differs slightly among the methods. The Metropolis method gives the largest peak and the Langevin equation with zero initial momentum gives the smallest peak. The reduced Langevin equation and the Langevin equation with thermal initial momentum distribution give almost the same result. In the previous calculation of the fission probability of the deformed nuclei, the left peak corresponds to the "ternary" (sequential binary) fission process and the right peak corresponds to the normal binary fission process. From the results, we can say that the use of the Metropolis method should be justified.

4 Summary

It is predicted that one of the fragment is extremely deformed in the binary fission of super-heavy nuclei. We expect that the deformed fragment may undergo the second fission thus producing three fragments. Starting from the pre-scission shape of the fragment of the first binary fission, we calculate the probability of the second fission by using the Metropolis method. Results are given for the fission of a super-heavy nucleus $^{300}120$ and a heavy actinide nucleus ^{252}Cf. The probability of the second fission is in the order of 10^{-4} for $^{300}120$ and 10^{-7} for ^{252}Cf, when we assume the constant temperature of $T = 1.0$ MeV. We also show that the ternary mass division is almost symmetric for the case of super-heavy nuclei. It should be mentioned that this work is just a simple, preliminary calculation of a new phenomenon that could be observed in the fission of super-heavy nuclei. More detailed and quantitative study is definitely necessary.

References

[1] R. Vandenbosh, J. Huizenga, *Nuclear Fission*, Academic Press (1973)
[2] V. Zagrebaev, A. Karpov, W. Greiner, Phys. Rev. C **81**, 044680 (2010)
[3] K.R. Vijayaraghavan and M. Balasubramaniam, W. von Oertzen, Phys. Rev. C **91**, 044616 (2015)
[4] Yu.V. Pyatkov, D.V. Kamanin *et al.*, Phys Rev C **96**, 064606 (2017)
[5] N. Carjan, F.A. Ivanyuk, Yu.Ts. Oganessian, Nucl. Phys. A **968**, 453 (2017)
[6] N. Carjan, F.A. Ivanyuk, Yu.Ts. Oganessian, G. Ter-Akopian, Nucl. Phys. A **942**, 97 (2015)
[7] V. Pashkevich, Nucl. Phys. A **169**, 275 (1971)
[8] M. Brack, J. Damgaard, A.S. Jensen, H.C. Pauli, V.M. Strutinsky, C.Y. Wong, Rev. Mod. Phys. **44**, 320 (1972)
[9] N. Metropolis, A.W. Rosenbluth, M.N. Rosenbluth, A.H. Teller, and E. Teller, J. Chem. Phys. **26**, 1087 (1953)

High resolution gamma-ray spectroscopy of fast-neutron induced fission

Jonathan Wilson[1,*], *Damien* Thisse[1], *Matthieu* Lebois[1], *Nikola* Jovancevic[1], *Rhiann* Canavan[2,3], *Mathias* Rudigier[2,4], *David* Étasse[5], and *Rosa-Belle* Gerst[6]

[1]Université Paris-Saclay, CNRS/IN2P3, IJCLab, 91405 Orsay, France
[2]Department of Physics, University of Surrey, Guildford GU2 7XH, UK
[3]National Physical Laboratory, Hampton Road, Teddington, Middlesex TW11 0LW, UK
[4]Technische Universität Darmstadt, Fachbereich Physik, Institut für Kernphysik, 64289 Darmstadt, Germany
[5]LPC Caen, 6 Boulevard Maréchal Juin, 14000 Caen, France
[6]Institut für Kernphysik, Universität zu Köln, 50937 Köln, Germany

Abstract. Experiments to perform precision spectroscopy of fast neutron induced fission were carried out during the ν-Ball experimental campaign at the ALTO facility of IJC Laboratory Orsay. Low energy fission of ^{232}Th(n,f), ^{238}U(n,f) and spontaneous fission of ^{252}Cf were studied using this hybrid high-resolution spectrometer and calorimeter. New observables such as γ-ray multiplicity distributions correlated with specific fission fragments are presented and discussed. A new method using fast-timing techniques to detect prompt fission neutrons in coincidence with prompt fission γ-rays is described.

1 Introduction

The ν-Ball experimental campaign was carried out in 2018 at the ALTO facility of the IJC Laboratory in Orsay. The ν-Ball spectrometer [1] is a hybrid Germanium, BGO and $LaBr_3$ spectrometer ideally suited to the study of fast-neutron induced fission.

The unique features of this device are as follows: (i) a hybrid spectrometer combining the high energy resolution of HPGe detectors with the excellent time resolution of $LaBr_3$ detectors; (ii) a fully digital acquisition system capable of running in triggered or triggerless modes; (iii) a high γ-ray detection efficiency; (iv) the capacity to perform γ-ray calorimetry due to the large solid angle coverage and (v) the ability to couple the spectrometer to the LICORNE directional neutron source [2], [3] to perform spectroscopy of neutron-induced reactions. The total Germanium photopeak efficiency was 6.7% at 1 MeV, and the total calorimeter photopeak efficiency was estimated at 48%.

The experimental campaign for fission studies took place over a period of 7 weeks to investigate fast neutron induced fission of ^{232}Th(n,f) at an average neutron energy of 1.7 MeV, ^{238}U(n,f) at average neutron energies of 1.7 and 3.5 MeV, and spontaneous fission of ^{252}Cf. The twin goals of these experiments were nuclear structure studies via high-resolution spectroscopy of neutron rich isotopes and studies of the fission process itself. The latter goal is the focus of this current paper.

*e-mail: jonathan.wilson@ijclab.in2p3.fr

1.1 Correlated fission observables

The fission observables accessible during the experiments were as follows:

- Prompt and delayed γ-ray multiplicity distributions
- Prompt and delayed γ-ray sum energy distributions
- Average prompt γ-ray multiplicities and total energies, $<M\gamma>$ and $<E_{tot}>$
- High energy γ-ray spectra up to 10 MeV
- Average prompt neutron multiplicities from fast timing (see Sect. 2)

which could all be correlated with the detection of individual fragments from observation of the characteristic prompt and delayed γ-ray decays in the HPGe detectors.

2 Prompt γ-ray multiplicity distributions

Prompt γ-ray multiplicity distributions are spin-sensitive observables that may contain interesting information which should be reproducible be fission models. These distributions correlated with individual fission fragments can be extracted by double-gating on γ-ray transitions in the HPGe detectors. Each multiplicity distribution will always have equal contributions from the nucleus in question and its binary partner fragments. It is not possible to separate the partial multiplicity contributions the two sources.

Currently only relative differences between these distributions are compared. To obtain absolute multiplicity distributions requires the application of a complicated unfolding procedure with the simulated ν-Ball response to γ-rays to events of different multiplicity and sum energies. The ν-Ball calorimeter is 68% efficient at detecting γ rays of 1 MeV. To obtain estimated absolute values for average multiplicities a scale factor of 1.3 can be applied.

2.1 Multiplicity distributions gating on spherical nuclei

The doubly magic nucleus ^{132}Sn is a special case when produced in fission, since it has a spherical shape and is very hard to deform. Its excitation spectrum is that of a vibrator rather than a rotor. In our previous work a large anomaly in the fission yield for this nucleus and its near neighbours was observed using measurement techniques of γ-ray spectroscopy [4]. Yields 500%–600% lower than expected were observed in the ^{238}U(n,f) reaction when compared to the available evaluations which are based on theory/extrapolations from other fissioning systems.

One explanation proposed for the origin of this anomaly was the low average γ-ray multiplicity, which would be theoretically emitted from ^{132}Sn due to its spherical shape at scission [5]. A lower multiplicity distribution associated with detection of this nucleus might cause a significant trigger bias and subsequent underestimation of its yield.

Figure 1 shows the coincident multiplicity distributions from the recent ν-Ball experiments for ^{132}Sn and its 3n partner nucleus ^{104}Mo. The shape of these distributions is observed to be very similar to that of other isotopes with a long tail extending to high multiplicities (> 10). This suggests that the explanation given in [5] cannot be the origin of the yield anomaly. Since ^{132}Sn is always emitted with a Mo partner nucleus, it is possible that a deficit of γ-rays from ^{132}Sn are compensated by a surplus of γ-rays from the partner.

Subsequent experimental work on fission yields of ^{238}U(n,f) also using the γ-ray spectroscopy technique by Fotiades *et al.* [6] shows that experimentally deduced normalized yields for Sn isotopes 3 times larger if only single γ rays are used to identify the fragments, instead of γ-γ coincidences (used in [4]). A more recent experiment by Ramos *et al.* [7] measured yields for ^{238}U(n,f) using the surrogate ^{9}Be (^{238}U, ^{239}U*)^{8}Be reaction which allows the possibility for direct identification of the fragments due to the inverse kinematics. Here, normalized fission yields were observed to be double those measured in [6], although it is not clear what the impact of the significant angular momentum mismatch between the surrogate and neutron-induced reactions had on the results. The remaining difference could possibility be due to a significant 0+ feeding, which was unobservable in [6].

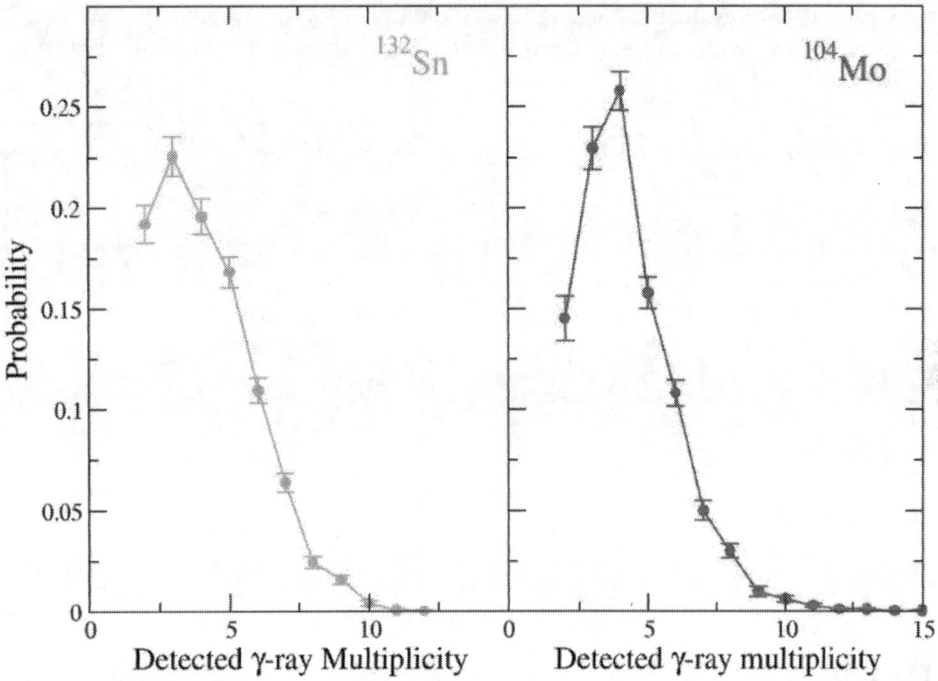

Figure 1. Comparison of the detected γ-ray multiplicity distributions for ^{132}Sn and ^{104}Mo from the ^{238}U(n,f) data. Fragments are identified by double-gating on the $4^+ \longrightarrow 2^+$ and $2^+ \longrightarrow 0^+$ transitions in the HPGe detectors.

A significant part of the fission yield anomaly is clearly due to the explanation suggested by Fotiades *et al.*, that for ^{132}Sn and its neighbours there will be a large direct side-feeding of the 2+ states. A lowering of the measured intensity occurs when using γ-γ coincidence spectroscopy to detect the isotope since the 2+ side-feeding is missed, but the effect is eliminated by using single γ ray detection of the $2{+}^+ \longrightarrow 0^+$ transitions. This explanation is confirmed by our current data, where direct side-feedings can now be measured by gating on the partner nucleus and measuring $4^+ \longrightarrow 2^+$ and $2^+ \longrightarrow 0^+$ ratios in the spectrum. The direct side-feedings are large and found to be large (up to 60%) in the case of ^{132}Sn and neighbours, while small for other nuclei.

2.2 Average multiplicity dependence on fragment mass

The evolution of average prompt γ-ray multiplicities as a function of the mass of the fragments may give interesting information, which ought to be reproducible by fission models.

In Fig. 2, we show the average multiplicity values deduced from double-gating on many different fragments produced in the ^{252}Cf reaction. Here, fission events were tagged directly by detection in an ionisation chamber placed in the centre of the array [8]. One fragment is stopped, while the other is detected in flight. The results show that the average multiplicity values are remarkably similar with very little variation as a function of mass. Fits to the data of two quadratic function in the light and heavy peaks show are clearly symmetric with a slight convex curvature towards fragments at peak yields, where average fragment multiplicity is ~10% higher. The average multiplicities always have two equal components from both the light and heavy peak partners, which are not possible to separate. Hence no saw-tooth patterns [9], [10] are evident due to high γ-ray multiplicity in one fragment summing with low multiplicity in the other fragment an vice-versa, resulting in a total multiplicity which is fairly constant.

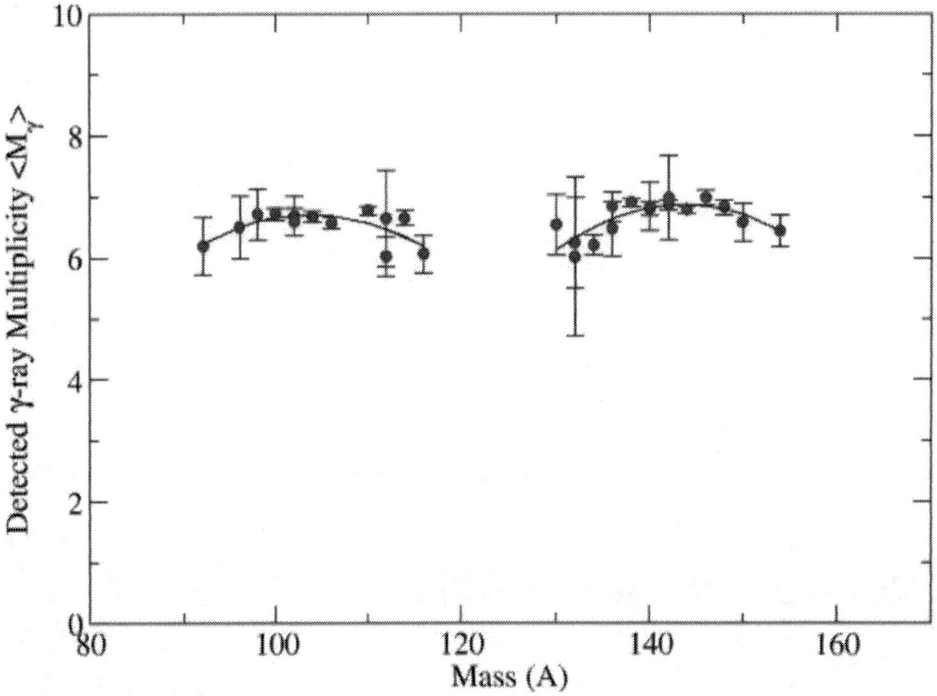

Figure 2. Average detected prompt γ-ray multiplicities for a range of even-even fragments produced in the ^{252}Cf(SF) reaction. Each fragment is identified by double gating on its $4^+ \longrightarrow 2^+$ and $2^+ \longrightarrow 0^+$ transitions in the HPGe detectors. Quadratic fits to the experimental data in heavy and light peaks are shown with the black lines.

2.3 Comparison of multiplicity distributions between the three systems

Multiplicity distributions can also be compared across all three of the fissioning systems. Figure 3 shows an example of the distributions for the ^{144}Ba nucleus which is close to the centre of the heavy peak for all three systems. The most frequently produced partner nuclei of ^{144}Ba are ^{104}Mo, ^{92}Kr and ^{86}Se for ^{252}Cf(SF), ^{238}U(n,f) and ^{232}Th(n,f) respectively. Large differences in multiplicity can be observed indicating a large drop in multiplicity as the fissioning system becomes lighter.

3 Correlated detection of prompt neutrons using fast TOF

The capacity to detect prompt fission neutrons in coincidence with prompt fission γ rays opens up new possibilities to investigate as-yet unexplored correlations between neutrons and γ rays. The ν-Ball spectrometer has this capability due to the presence of the $LaBr_3$ detectors in two rings at distances of 11 cm and 15 cm from the fission target. Since the neutron beam is pulsed with a period of 400 ns and a pulse width of ~2 ns, prompt neutrons can be detected via time-of-flight (TOF) from inelastic scattering in the $LaBr_3$ detectors. Figure 4 shows the two dimensional separation of prompt γ ray and prompt neutron events in the time-energy plane. Neutrons in the energy range 0.165–2.55 MeV can be detected in the ring at a distance of 11 cm and 0.165–4.75 MeV can be detected in the second ring at 15 cm corresponding to neutron flight times between 5 and 35 ns. To be certain the delayed γ-ray detection originated from inelastic scattering of a prompt neutron in the detector an additional condition is placed on the γ ray energy. Energies corresponding to the first or second excited states of ^{139}La, ^{79,81}Br must be detected at 165, 217, 261 and 276 keV (see Fig. 4). The integral of spectral counts is proportional to the number of neutrons detected giving a total efficiency for neutron detection of $\bar{\varepsilon} = 0.15\%$. This efficiency is calibrated using the ^{252}Cf fission tagged data and the known ^{252}Cf neutron multiplicity of $\bar{\nu} = 3.73$ per tagged fission event. A more sophisticated extension of the technique would be to weight each neutron according to the inelastic scattering cross section at the detected energy on an event-by-event basis.

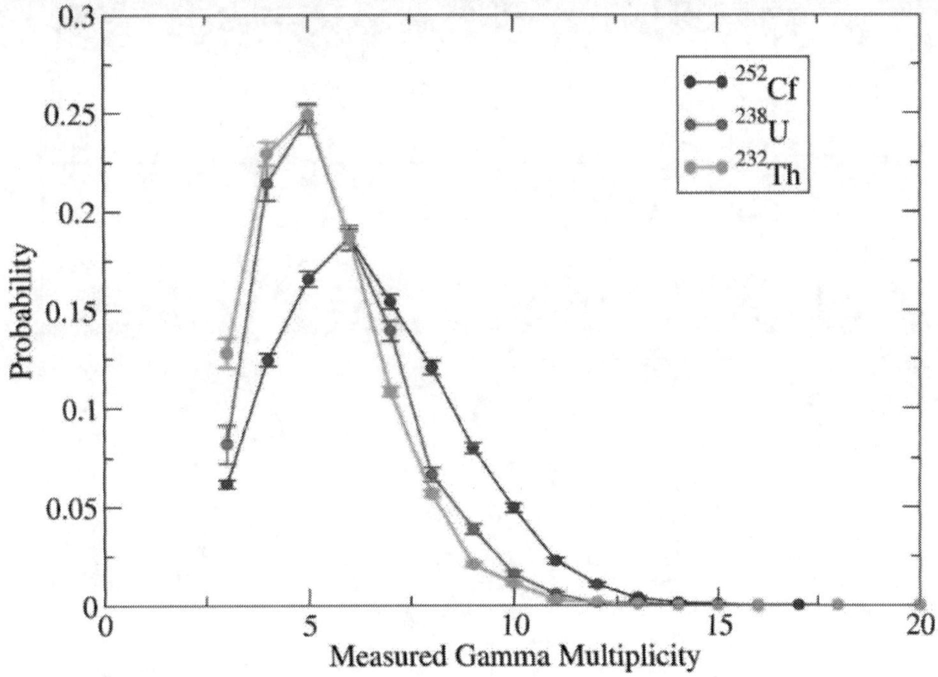

Figure 3. Comparison of the detected prompt γ-ray multiplicity distributions for the ^{144}Ba nucleus for the 3 different systems. Identification of ^{144}Ba is carried out by double-gating on the $4^+ \longrightarrow 2^+$ and $2^+ \longrightarrow 0^+$ transitions in the HPGe detectors.

For the neutron-induced fission data, care has to be taken to subtract off an appropriate background of neutrons uncorrelated in time due to scatterings occurring into the spectrometer from the neutron source itself. This problem does not occur for the ^{252}Cf spontaneous fission data. The capacity to detect prompt neutrons allows variations in prompt fission neutron multiplicity to be measured. Work to explore these neutron/γ correlations in fission is ongoing.

4 Conclusions

The ν-Ball spectrometer coupled to the LICORNE neutron source offers some interesting new possibilities for deeper investigations into the fission process.

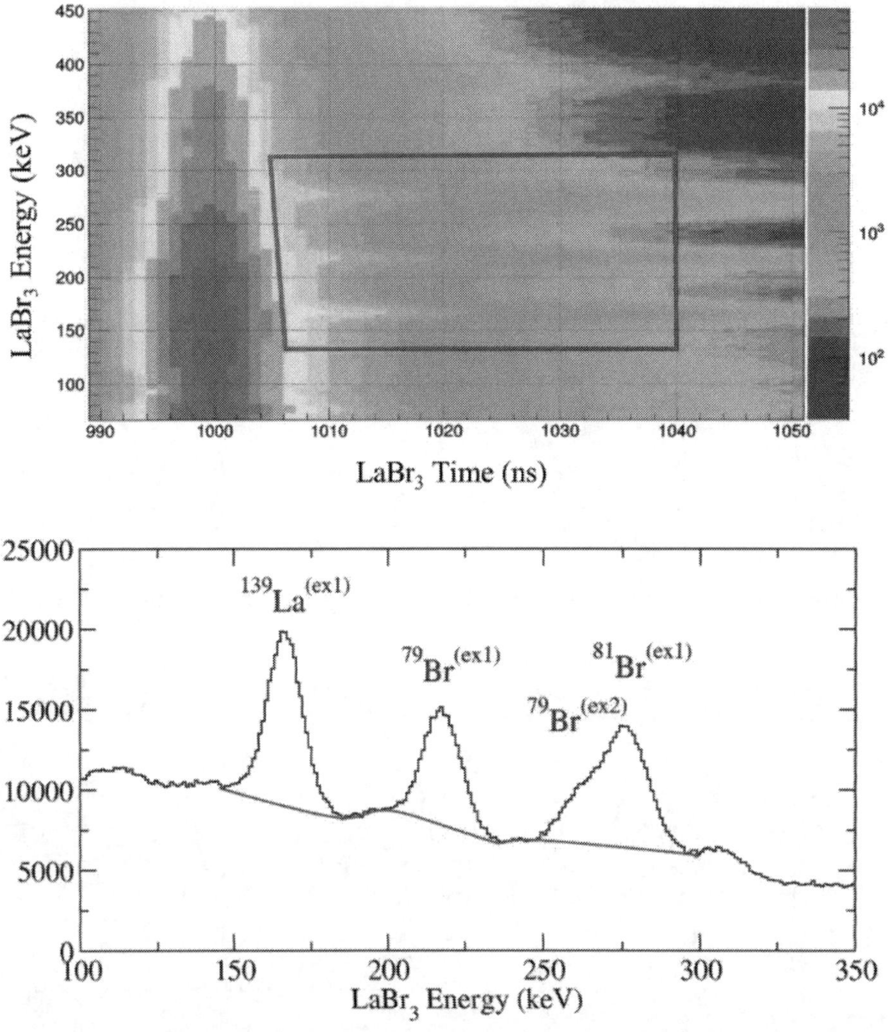

Figure 4. Detection of delayed γ rays from inelastic scattering of prompt neutrons in the $LaBr_3$ detectors. The top panel shows the 2-dimensional separation in the time-energy plane, while the bottom panel shows the projection onto the energy axis of the selected red zone and integration of the relevant peaks.

In particular, new observables such as prompt γ-ray multiplicity distributions are accessible in coincidence with particular fission fragments. Detection of prompt neutrons via time-of-flight over short distances is possible and variations in prompt neutron multiplicity can be measured. The complex correlations between these various observables can now be explored and analysis work is ongoing.

References

[1] M. Lebois, J.N. Wilson, et al., Nucl. Instrum. Meth. A **735**, 145 (2014)
[2] J.N. Wilson, M. Lebois, M., et al., Phys. Proc. **64**, 107 (2015)
[3] M. Lebois, N. Jovancevic, D. Thisse, R. Canavan, D. Étasse, M. Rudigier, J.N. Wilson, Nucl. Instrum. Meth. Phys. Res. A **960**, 163580 (2020)
[4] J.N. Wilson et al., Phys. Rev. Lett. **118**, 222501 (2017)
[5] H. Pasca et al., Phys. Rev. C **98**, 014624 (2018)
[6] N. Fotiades et al., Phys. Rev. C **99**, 024606 (2019)
[7] D. Ramos et al., Phys. Rev. Lett. **123**, 092503 (2019)
[8] L. Gaudefroy, et al., Phys. Rev. C **97**, 064317 (2018)
[9] J.N. Wilson et al., Nature **590**, 566–570 (2021)
[10] M. Travar et al., Phys. Lett. B **817**, 136293 (2021)

AUTHOR INDEX